The Chemistry and Microbiology of Pollution

The Chemistry and Microbiology of Pollution

I. J. HIGGINS and R. G. BURNS
University of Kent at Canterbury, England

1975

ACADEMIC PRESS
London New York San Francisco

A Subsidiary of Harcourt Brace Jovanovich, Publishers

ACADEMIC PRESS INC. (LONDON) LTD.
24–28 Oval Road
London NW1

U.S. Edition published by
ACADEMIC PRESS INC.
111 Fifth Avenue
New York, New York 10003

Library of Congress Catalog Card Number: 75-19649
ISBN: 0-12-347950-9

Printed in Great Britain by William Clowes & Sons Limited
London, Colchester and Beccles

Preface

Despite all that has been said and written in recent years about pollution and the environmental crisis, there has been a conspicuous lack of a comprehensive text dealing with the chemistry and microbiology of the subject. It is this deficiency that has prompted the writing of this book, which is aimed at both undergraduates and graduates studying environmental sciences. We hope that thc specialist will also find the book useful even though he may discover insufficient detail in some areas; this must be inevitable in a book of this size covering such a variety of topics. Reference is made, where appropriate, to more comprehensive texts.

There is no attempt at a detailed discussion of the moral and philosophical aspects of environmental pollution which have been argued at great length elsewhere. We hope, however, that the reader will find in this volume a balanced view of the current state of chemical and microbiological knowledge and that the gaps in mankind's understanding will be apparent to him. These must be filled by further research if we are to adopt a rational approach to the increasingly serious pollution problems resulting from a rapidly expanding population and a concomitant desire for higher standards of living.

Finally, we should like to thank Dr. R. B. Cain for access to his review on surfactant degradation prior to its publication, our patient wives for typing and proof-reading, the publishers for many helpful suggestions and Dr. Hugo Z. Hackenbush for keeping us sane throughout the past months.

University of Kent,
July, 1975

I. J. Higgins,
R. G. Burns

Contents

1
Introduction

The biosphere is dominated by a reactive mixture of hydrogen, carbon, oxygen, nitrogen, phosphorus and sulphur compounds—all in a continuous state of synthesis, decomposition and transformation. It is upon this dynamic equilibrium that the success or failure of our ecosystem depends. The functions of microorganisms in these processes include fixation, assimilation, and the degradation of organic residues to release nutrients in a suitable form for uptake by plants. This last process is described as mineralization and is a key reductive step in the recycling of elements.

The apparently unfailing capacity of bacteria to mineralize naturally occurring organic matter was described by Beijerink as "microbial infallibility". To a large extent man-made organic substances are also degraded by microorganisms although, in some instances, the time required for total breakdown is considerable. However, some synthetic chemicals appear resistant to both biological and non-biological decay, representing not only the immobilization of elements but also the accumulation of toxins.

Some ecologists believe that a pollutant-induced disruption of biological cycles may lead to irreversible damage to the environment. It is probable, however, that all ecosystems have some degree of flexibility and can absorb quite severe shocks without permanent displacement of their equilibria. Notwithstanding, some toxic compounds are not only recalcitrant but also exhibit mobility within the biosphere; a combination which may have undesirable ecological consequences.

1. Origin and Dispersal of Pollutants

Environmental pollutants, defined as those chemicals having a detrimental effect on man and his environment, are derived from three main sources (Table 1.1).

TABLE 1.1
Origin of pollutants

A. Naturally occurring	B. Transformed and Concentrated	C. Synthesized
Oxides of nitrogen	Sewage	Pesticides
Nitrate	Fertilizers	Surfactants
Nitrite	Acid Waste	Radionuclides
Asbestos	Fuel combustion products	Synthetic polymers
Heavy metals	Heavy metals	Petrochemicals
Radionuclides	Radionuclides	
Hydrocarbons and their derivatives	Pesticides	
	Surfactants	
Allergens	Hydrocarbons	
	Petrochemicals	

In the first instance they may arise quite naturally to form part of the background level of toxic substances in the environment. Constant distribution and genetic selection has ensured that these compounds cause few problems and even if absorbed by plants and animals are rapidly excreted or detoxified. Pollutants may also be produced by the concentration and transformation of naturally occurring materials during their domestic and industrial use. Finally, some novel chemicals are synthesized by man.

From their point of origin pollutants may spread laterally and vertically to all components of the environment and their persistence can be measured in minutes (sulphur dioxide), days (alkoxy-alkyl-mercury), years (DDT) or even centuries (plutonium-239). The capacity of a chemical to persist and become a pollutant rests upon its inherent physico-chemical properties, its resistance to removal by chemical, physical and biological mechanisms and its toxicity to microorganisms, plants and animals.

2. Environmental Problems

It has been said that mankind's concern for his environment is directly related to the ability of the chemist to detect low levels of toxic chemicals. To some extent this is true, in the sense that our understanding of the consequences of contaminating the biosphere has somewhat paralleled the development of analytical techniques. Nevertheless, the dangers of pollutants to man and his environment are real enough (Table 1.2) and we are all familiar with phenomena such as eutrophication, during which aquatic systems are deoxygenated by the introduction of reduced effluents and inorganic

TABLE 1.2

The effect of pollutants on man

Pollutants may affect man by:

1. his inhalation of contaminated air
2. his ingestion of contaminated food and water
3. damaging his livestock and crops
4. increasing levels of radiation
5. causing climatic changes
6. inhibiting the flora and fauna and disrupting biological stability
7. affecting microorganisms, plants and animals responsible for mineral cycles
8. disrupting amenities

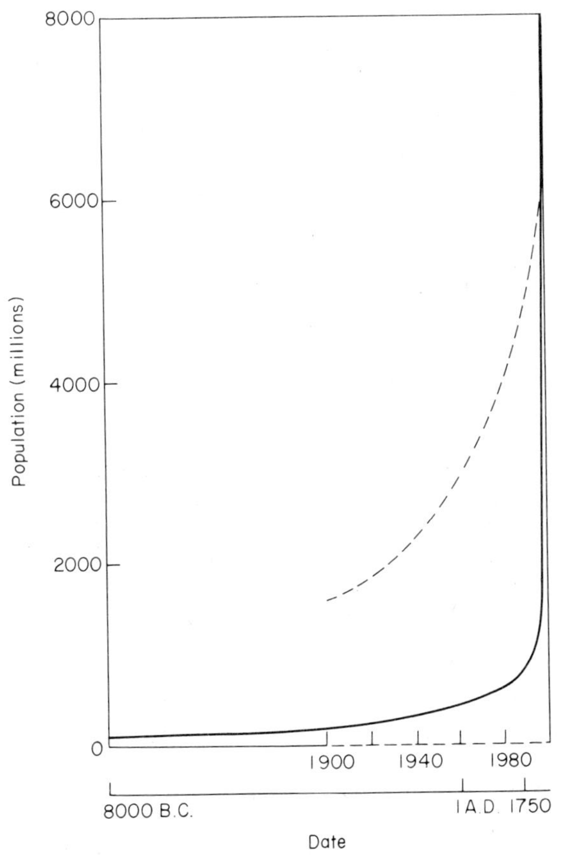

Fig. 1.1. Human population growth.

nutrients. Additionally, DDT, cyclodiene insecticides, mercury and cadmium may become concentrated in animal tissues and cause the disruption of a variety of normal metabolic and physiological functions whilst the hazards of air pollution have been dramatically demonstrated.

On the other hand, over-reaction has resulted in the total condemnation of many chemicals before any assessment of real value versus real or potential dangers has been made. In recent years a highly critical and somewhat emotional spotlight has been directed at phosphates, methyl mercury, nitrate, organo-chlorine insecticides and others

In the broadest sense it is difficult to divorce pollution problems from those posed by a rapidly expanding population (Fig. 1.1). In the simplest terms this means more people: more food; an equation that is becoming increasingly difficult to balance. It has been estimated that some 50% of the world's potential agricultural land is already being used and that the cultivation of the remainder will prove extremely difficult. The potential agricultural acreage will, of course, decrease with population growth and, at present productivity levels and predicted population demands, will have ceased to produce enough food by the end of the century. Some have forecast that increasing the yield four-fold (fertilizers, pesticides, genetic manipulation) will only delay the nutritional catastrophe for two generations. Additionally, the greater material expectations and energy consumption of many societies puts pressure upon industries whose products and by-products contribute significantly to pollution.

It is readily apparent that we need to (i) improve our understanding of both the short- and long-term effects of pollutants and (ii) investigate alternative technologies (e.g. biological control). The former will only result from a comprehensive understanding of the molecular structure of pollutants, their chemical and physical properties, their mobility and partitioning in the hydrosphere, atmosphere, lithosphere and biosphere, and their rate of and susceptibility to biological and non-biological decay.

Recommended Reading

Articles on Human Population. (1974). *Scientific American* 231 No. 3. W. H. Freeman.

Bourne, A. (1972). "Pollute and Be Damned." Dent & Sons. London.

Carson, R. L. (1962). "Silent Spring." Houghton Mifflin Co., Boston.

Chemistry in the Environment. (1973). Readings from the *Scientific American*, W. H. Freeman.

Detwyler, T. R. (1971). "Man's Impact on Environment." McGraw-Hill.
Edwards, R. W. (1972). "Pollution." Oxford Biology Reader No. 31. O.U.P.
Erlich, P. R. and Erlich, A. H. (1970). "Population, resources, environment-issues in human ecology." W. H. Freeman.
Mellanby, K. (1967). "Pesticides and Pollution." New Naturalist Series. Fontana.
Mitchell, R. (1974). "Introduction to Environmental Microbiology." Prentice Hall, New Jersey.
Strobbe, M. A. (Ed.) (1971). "Understanding Environmental Pollution." C. V. Mosby Co., St. Louis, Mo.

2
Pesticides

INTRODUCTION

Pesticides can be defined as those chemicals employed by Man to destroy or to inhibit life forms which he has decided are a nuisance. Of course, this may often be a subjective assessment in that one man's weed is another's wild flower.

Used as a collective noun, the word pesticide embraces herbicides, fungicides, insecticides, nematocides, rodenticides, molluscicides, bacteriocides and others (Fig. 2.1). Compounds which have been and which are being used for biological control include naturally occurring organics (antibiotics, pyrethrins), synthetic organics (chlorinated hydrocarbons, thiocarbamates) and inorganics (copper sulphate, mercuric chloride). Since 1945 the production and use of organic pesticides in particular, has increased at an enormous rate (for example see Table 2.1) and the prominence of the subject in the public mind is, to a large extent, due to any associated environmental hazards.

Pesticide pollution is not, however, a new problem. Prior to the Second World War pesticides were inorganic, made up of arsenic, mercury, selenium and lead compounds amongst others and routinely applied at dosage rates in excess of 200 kg ha^{-1}. Even at

Fungicides		Fungi
Insecticides		Insects
Herbicides		Green Plants
Nematacides	INHIBIT OR KILL	Nematodes
Molluscicides		Molluscs
Bacteriocides		Bacteria
Rodenticides		Rodents

Fig. 2.1. What are pesticides?

TABLE 2.1

Production of organochlorine insecticides in the U.S.A.

	kg x 10^{-6}	
	DDT	Cyclodienes
1945	15.9	
1950	34.1	
1955	59.1	36.4
1960	75.0	43.2
1965	63.6	52.4
1969	60.5	50.0

this time there was some concern over the accumulation of insoluble heavy metal residues in soil (see Chapter 7). However, in the postwar period, with the introduction of organic herbicides such as 2,4-dichlorophenoxyacetic acid (2,4-D) and the chlorinated hydrocarbon insecticides like DDT, the use of inorganic chemicals for pest control declined rapidly. In addition, the whole approach to pest control was revolutionized with the development of highly selective and efficient compounds which could be used at low application rates (1–20 kg ha^{-1}). Moreover, it was hoped that, since organic compounds are readily metabolized by microorganisms, toxic residue problems would be negligible. Unfortunately, it has since become apparent that microorganisms are far from infallible in their ability to degrade each and every organic substrate presented to them. As a result, whilst most organic pesticides are transient some may remain unchanged in the environment for long periods of time (1–20 years) whilst others may be converted to equally persistent toxic substances. In the early 1950's environmentalists were beginning to voice their concern over the apparent persistence of these "biodegradable" organic compounds and, as a group, the chlorinated hydrocarbon insecticides (DDT, aldrin, dieldrin, heptachlor) were the subject for many of these misgivings.

It was not, however, until the early 1960's that the public at large became aware of the deleterious effects of pesticide residues in the environment. Around this time Rachel Carson, a scientific and literary figure of some repute, published "Silent Spring" a book in which she dramatically described the catastrophic effects on wild life arising from the indiscriminate use of pesticides. The remarkable outcome of this book was that, in addition to becoming a best seller, it focused public attention on the potential problems associated with pesticide use, specifically in relation to non-target animals.

CHEMISTRY

Pesticides include a vast array of substances with widely different formulae and chemical and physical properties. In this discussion only the major groups of compounds are introduced and the reader is referred to the bibliography at the end of the chapter for a more detailed account.

On a chemical basis pesticides may be conveniently divided into three major groups: inorganic compounds, organic compounds containing metal ions, and organic compounds.

1. Inorganic Pesticides

1.1. Insecticides

Inorganic arsenic compounds, in addition to their use as herbicides, have also found favour as non-selective insecticides. The first of these compounds was Paris Green, ($Cu(C_2H_3O_2)_2 . 3Cu(AsO_2)_2$) introduced in 1865 to control the Colorado Potato Beetle. Probably the most widely used and the safest of the arsenicals since that time has been lead arsenate ($PbHAsO_4$), traditionally mixed with sodium arsenate and suitable for controlling a wide spectrum of insect pests. Compounds of fluorine have been utilized as insecticides in the past but have not proved very effective.

1.2. Fungicides

Salts of copper, zinc, mercury, potassium and sulphur have been used as fungicides for many years. In the second half of the eighteenth century solutions of copper sulphate ($CuSO_4 . 5H_2O$) were used to protect wheat seed and wood from fungal attack. Bordeaux mixture, a formulation of copper sulphate and calcium carbonate, was introduced in 1885 and, as an effective mildew-controlling fungicide, is still in use today. Zinc oxide (ZnO) is commonly used in fungicide preparations and has proved particularly useful in the treatment of superficial mycoses such as Athlete's Foot.

Mercuric chloride ($HgCl_2$) was introduced in 1890 as a fungicide for treating cereal seeds and had been used prior to this as a bacteriocide. However, due to its chronic and acute mammalian toxicity, it was never widely adopted in agricultural practice. Mercuric cyanide ($Hg(CN)_2$) or corrosive sublimate, however, has been used in the treatment of fungal and insect pathogens of potatoes, cabbages, cucumbers and fruit trees although it is also extremely poisonous to Man.

Potassium sulphide solutions are effective substitutes for Bordeaux mixture in the control of powdery mildews whilst sulphur and sulphur-lime have been used to alleviate fungal infections of stone fruit (such as peaches and plums) and as insecticides for cattle dips.

1.3. Herbicides

A variety of inorganic compounds are used as herbicides. Arsenic trioxide (As_2O_3), sodium arsenite ($NaAsO_2$) and calcium arsenate ($Ca_3(AsO_4)_2$) have been applied as soil sterilants and non-selective contact and systemic herbicides. However, in recent years, they have been largely replaced by organic arsenicals which are more selective to plants and, at the same time, less toxic to mammals.

Sodium chlorate ($NaClO_4$) is widely used for the destruction of deep-rooted perennials. This chemical is a strong oxidizing agent, a useful characteristic in the firework industry but one presenting a not inconsiderable hazard when used as an herbicide. Consequently sodium chlorate is often mixed with various borate salts and substituted ureas, both to decrease its combustibility and increase its weed-killing efficiency.

Boron compounds (as sodium tetraborate or borax, $(Na_2B_4)_7 . 10H_2O$) are extremely toxic to plants and may persist in the soil for a considerable time. Unlike sodium chlorate however borax herbicides are non-flammable, non-corrosive, non-volatile and non-poisonous.

Calcium cyanamide ($CaCN_2$) is a fertilizer, defoliant and herbicide. Other inorganic herbicides include sodium cyanate (NaOCN), potassium cyanate (KOCN), ammonium sulphamate ($H_2NS(O)_2ONH_4$) and magnesium chlorate ($Mg(ClO_3)_2 . 6H_2O$).

1.4. Fumigants

Several simple inorganic compounds, such as carbon bisulphide (CS_2), carbon tetrachloride (CCl_4), hydrogen cyanide (HCN) and sulphur dioxide (SO_2), function as fumigants to destroy a variety of pests.

2. Organo-Metallic Pesticides

The dangers involved in handling mercuric chloride prompted the introduction in 1915 of chlorphenol mercury ($ClC_6H_3OH . Hg . OSO_3Na$), the first of the organic mercurial pesticides. Subsequently a variety of compounds were synthesized all having the general formula of RHgX, where R = aryl-, aryloxy-, alkyl-, alkyloxyethyl-, and X = chloride, acetate, lactate, urea of hydroxyl. Examples include methoxyethylmercury chloride ($CH_3O \cdot CH_2 \cdot CH_2 \cdot HgCl$), hydroxymercurychlorophenol ($Cl \cdot C_6H_4 \cdot O \cdot HgOH$), phenylmercuric acetate ($C_6H_5Hg \cdot O \cdot CO \cdot CH_3$) and ethylmercuric chloride ($C_2H_5 \cdot HgCl$). Most of these substances have been used as seed treatments for fungal control for once in the soil they decompose to yield the active fungicidal agent, either metallic mercury or a mercury salt. Unfortunately mercury compounds are rapidly absorbed through the skin, mouth and respiratory tract and show both an acute and a chronic toxicity to mammals. These pesticides also pose a broader environmental hazard, since they are concentrated in food chains as biologically-formed dimethylmercury (see Chapter 7).

Since 1950 a series of organo-tin compounds has been developed and, although as a group they tend to be rather general biocides, they are safer fungicides than the organo-mercury compounds. The organo-tins include tributyl-tin hydroxide ($(C_4H_9)_3SnOH$), triphenyl-tin acetate ($(C_6H_5)_3SnOCOCH_3$) and triphenyl-tin disulphide ($(C_6H_5)_3SnS{-}SSn(C_6H_5)_3$). Triphenyl-tin compounds have the lowest toxicity to higher plants of the derivatives so far synthesized and can be used as foliar fungicides.

Cacodylic acid, monosodium methane-arsonate and disodium methane-arsonate are all arsenic-containing organic herbicides.

3. Organic Pesticides

This group accounts for the majority of pesticides in present day use.

3.1. Insecticides

Table 2.2 presents the formulae of a variety of organic insecticides.

3.1.1. Naturally Occurring Compounds. Nicotine, in the form of dried tobacco leaves and stems, was employed to control aphids on currants and gooseberries as long ago as 1763. It was thus one of the earliest organic pesticides and is still used today. It is, however, acutely toxic to Man and other animals but since it is rapidly broken down is not considered a chronic hazard.

The main pyrethroid insecticides are synthetic analogues of the naturally occurring pyrethrins from *Chrysanthemum* species. They have extremely low mammalian toxicities since they are rapidly metabolized in the mammalian body. Insects are also capable of

TABLE 2.2
Organic insecticides

Chemical Group	Examples	Formula
Nicotinoids	Nicotine	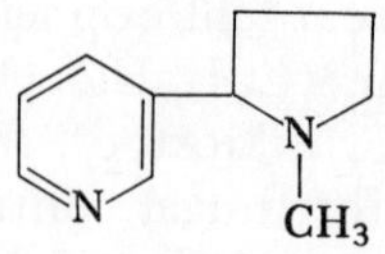
	Anabasine	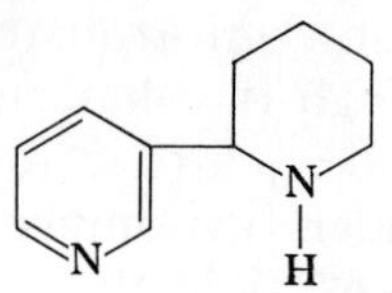
Pyrethoids	Pyrethin 1	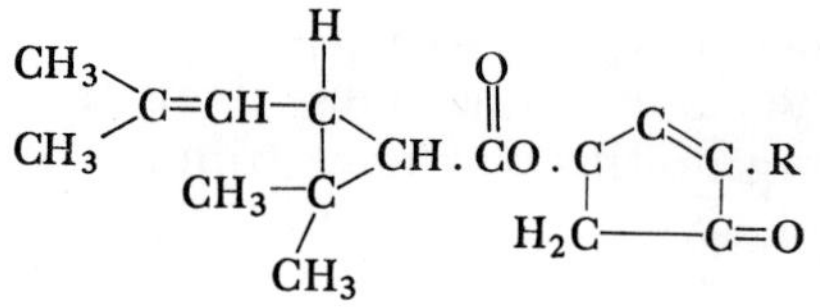 ($R = CH_2CH{=}CH.CH{=}CH_2$)
Rotenoids	Rotenone	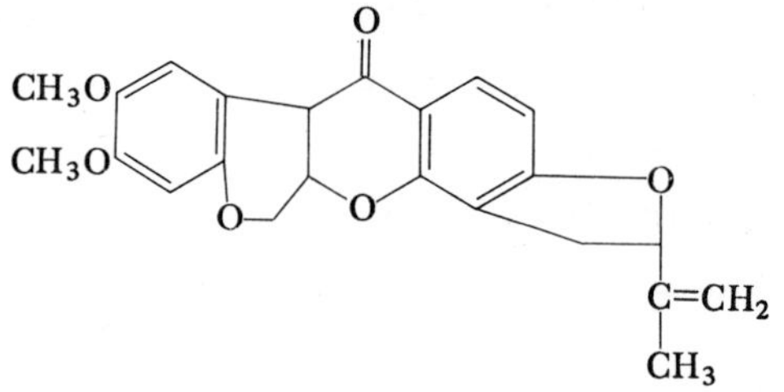
Dinitrophenols	DNOC	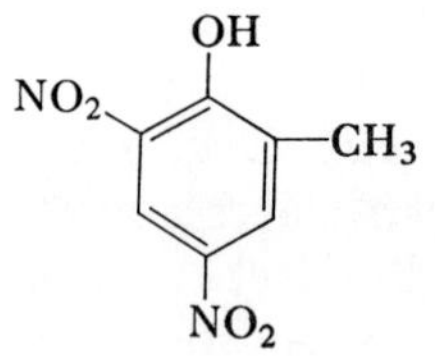
	DNOSBP	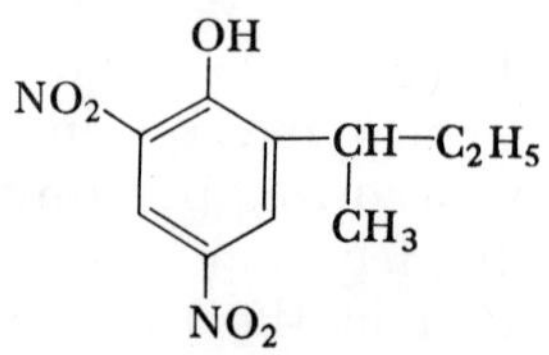

TABLE 2.2—*continued*

Chemical Group	Examples	Formula
Organothiocyanates	Thanite	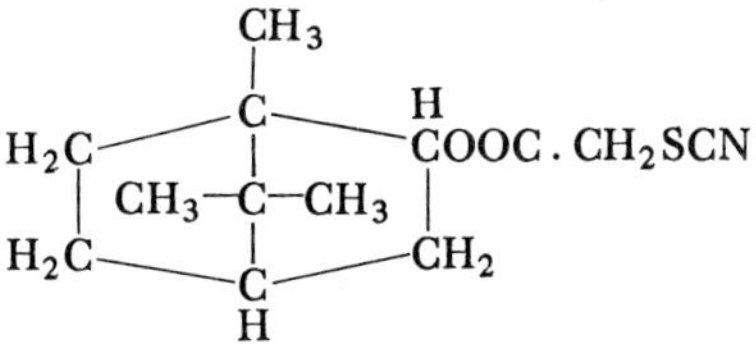 CH_3; H_2C; C; H; $COOC.CH_2SCN$; CH_3-C-CH_3; H_2C; C; CH_2; H
DDT and analogues	DDT	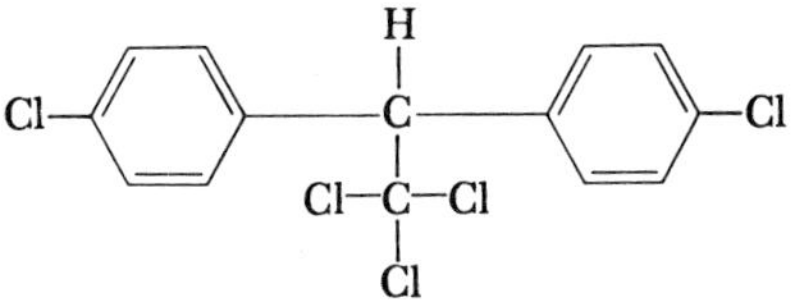 H; Cl; C; Cl; Cl–C–Cl; Cl
	Methoxychlor	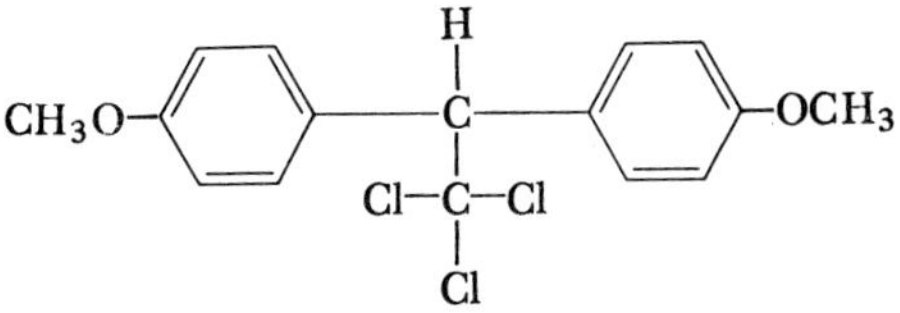 CH_3O; H; C; OCH_3; Cl–C–Cl; Cl
Cyclodienes	Chlordane	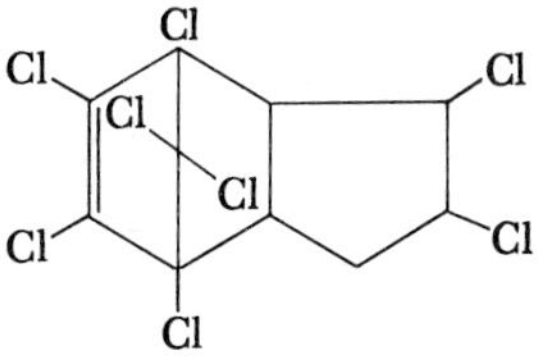 Cl; Cl; Cl; Cl; Cl; Cl; Cl; Cl
	Heptachlor	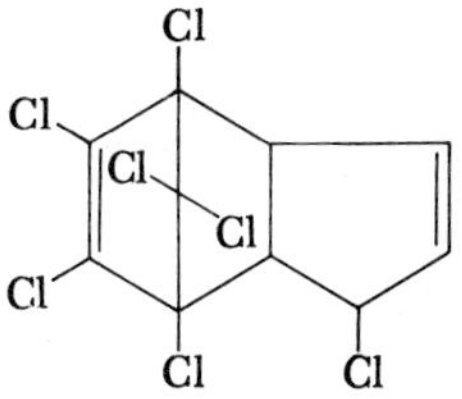 Cl; Cl; Cl; Cl; Cl; Cl; Cl
	Aldrin	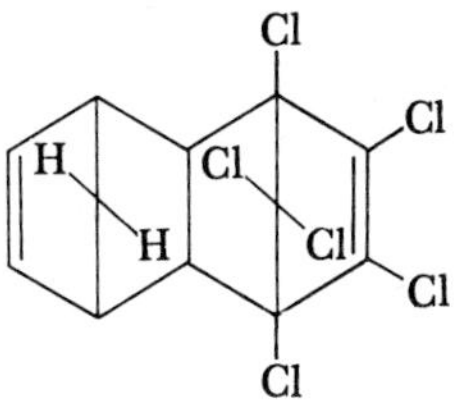 Cl; Cl; H; Cl; H; Cl; Cl; Cl

TABLE 2.2—*continued*

Chemical Group	Examples	Formula
	Dieldrin	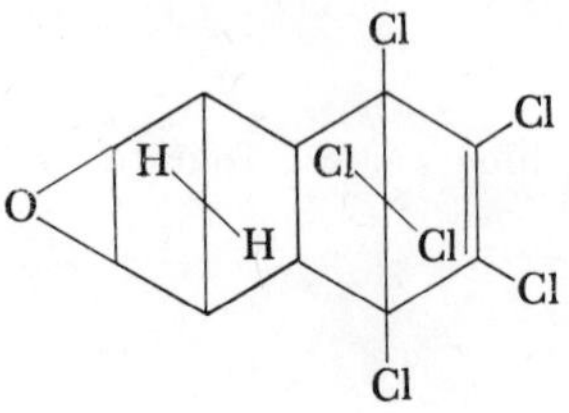
Carbamates	Carbaryl	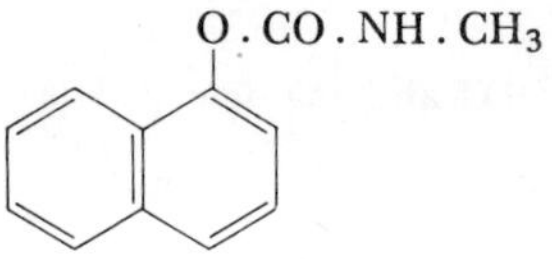
	Baygon	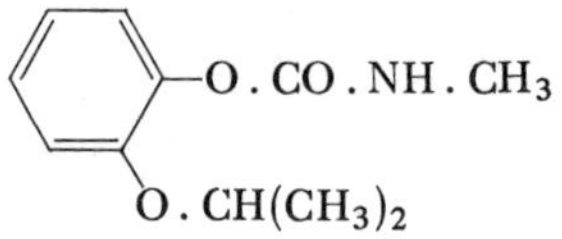

metabolizing these compounds which are often administered along with synergists which inhibit the detoxifying enzymes and, in consequence, potentiate the activity of the insecticide. These synergists are usually derivatives of methylenedioxybenzene (I)

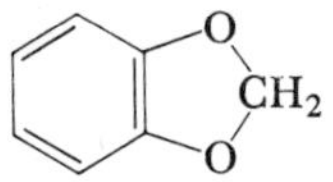

[I]

3.1.2. Dinitrophenols. Potassium dinitro-δ-cresylate was the first of the synthetic organic insecticides although the dinitrophenols, in general, are compounds with a much wider range of biocidal activities. In addition to their direct function as insecticides they function as acaricides, herbicides and fungicides and as ovicidal sprays for the control of mites, aphids and scale insects.

3.1.3. Thiocyanates. Lethane 384 was the first commercially employed synthetic organic insecticide and was marketed in 1932 as a rapid "knock down" agent for the control of household insects. Today the thiocyanates are still used principally in household aerosols because they act rapidly and have a low toxicity to higher animals.

3.1.4. Chlorinated Hydrocarbons. The chlorinated hydrocarbon group includes a number of important and widely used insecticides—such as DDT and its analogues and the chlorinated cyclodienes aldrin, dieldrin, chlordane and heptachlor. DDT was first synthesized in 1874 although its insecticidal properties were not discovered until some 65 years later. It has been, for many years, the most widely used insecticide with a world production in 1970 of about 72 000 tonnes. Its value lies in its potency, persistence and cheapness of manufacture. However, it is this very persistence, related to the fact that it accumulates in lipid tissues and as a consequence in food chains, that has been a cause of concern in recent years. The biomagnification of DDT is a subject discussed elsewhere in this chapter. Some insects have become resistant to DDT because they produce an enzyme DDTase, which converts DDT to the inactive DDE. In some cases the efficacy of DDT can be restored by applying it together with an analogue known as antiresistant (II)

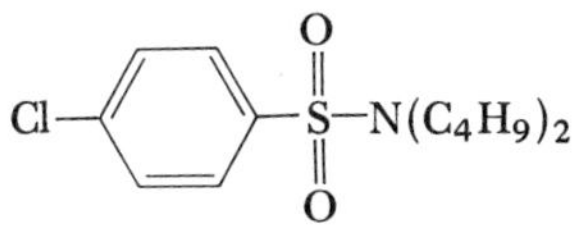

$$Cl-C_6H_4-S(=O)_2-N(C_4H_9)_2 \qquad [II]$$

Another analogue, methoxychlor, is also a very effective insecticide but unlike DDT it is not concentrated in animal fat.

Amongst the cyclodienes, chlordane and heptachlor are used chiefly to control cockroaches, ants, termites and soil insects. Aldrin and dieldrin are both broad spectrum insecticides and, of the two, dieldrin is the more stable whilst aldrin will quite readily break down to form dieldrin. Heptachlor and aldrin are rapidly oxidized in both plant and animal tissues to their more stable epoxides. For instance when heptachlor is fed to animals it is concentrated about twenty times and stored in fat tissues as heptachlor epoxide. This may lead to the same undesirable accumulation in food chains as described for DDT. This environmental characteristic is common to all the cyclodiene insecticides.

3.1.5. Organophosphorus Compounds. Although, on a strictly chemical basis, the organophosphorus compounds should be included amongst the organo-metallic pesticides, they are normally discussed within the context of organics. About forty organophosphorus compounds are commercially available and it is thought that in the region of 100 000 have been synthesized for assessment as insecticides. The general formula for these compounds, together with the specific formulae of the more common ones are shown in Table 2.3.

TABLE 2.3

Organophosphorus insecticides

General Formula: $R(R')P(=S(O))-X$

Name	Formula
Parathion	$NO_2-C_6H_4-O-P(=S)-(OC_2H_5)_2$
Methyl parathion	$NO_2-C_6H_4-O-P(=S)-(OCH_3)_2$
Dicapthon	$NO_2-C_6H_3(Cl)-O-P(=S)-(OCH_3)_2$
Chlorthion	$NO_2-C_6H_3(Cl)-O-P(=S)-(OCH_3)_2$
Malathion	$(CH_3O)_2P(=S)-S-CH(-COOC_2H_5)-CH_2-COOC_2H_5$
Phosdrin	$(CH_3O)_2 . P(=O) . OC(CH_3){=}CH . COOCH_3$
Thimet (Phorate)	$(C_2H_5O)_2 . P(=S) . SCH_2SC_2H_5$
Meta-systox	$(CH_3O)_2P(=O) . S . CH_2CH_2S(=O)C_2H_5$

The most widely used are parathion and its analogue methyl-parathion. Parathion itself is highly toxic to mammals, whilst the methyl analogue is rather less toxic but more effective against aphids and beetles. The chlorinated derivatives, dicapthon and chlorthion, are used as domestic insecticides because they show little mammalian toxicity. Malathion is an apparently safe general purpose insecticide for use in house and garden.

Although all these insecticides function in much the same way (in that they inhibit the enzyme acetylcholine esterase essential for normal transmission of nerve impulses) their relative toxicity to insects and mammals varies a great deal. For example, malathion is between 5 and 50% as toxic to insects as parathion but only 0.5% as toxic to mammals. The explanation for this is that malathion is rapidly detoxified in the mammalian liver but not in the insect body. Dicapthon and chlorthion also have low mammalian toxicity but are as toxic to insects as parathion.

Several organophosphorus compounds are used as systemic insecticides, particularly phorate and disulphoton. These two are not themselves active but are oxidized in the plant tissues to sulphoxide and sulphone metabolites which then function as insecticides.

3.1.6. Miscellaneous. Petroleum oils are used as pesticides in dual roles; firstly as dispersants and secondly as the actual toxicants. In general, higher boiling point fractions are best as insecticides while more aromatic fractions are used as herbicides. The insecticidal activity of oils is due to a suffocating effect and it seems that the higher boiling fractions are more efficient in this regard, because of their higher viscosity. The herbicidal activities of petroleum oils are complex and poorly understood.

Several microbial products have found favour as insecticides and are notably both highly specific and non-toxic to higher animals. For example, *Bacillus thuringiensis* contains a crystalline, proteinaceous toxin that has been used to control the cabbage looper and the alfalfa caterpillar.

Compounds used for mothproofing include paradichlorobenzene and naphthalene.

3.2. Fungicides

The fungicides fall into six major groups of chemicals (see Table 2.4).

3.2.1. Chlorinated Phenols. The chlorinated phenols are used as wood preservatives and to prevent the formation of microbial slimes in the paper mill industry. Chlorination of simple phenols increases their fungitoxicity which is maximal with three or five chlorine atoms. Bis-chlorophenols, in which two rings are joined by a methylene bridge (dichlorophene) or a sulphur atom (Vancide BL), are especially good for mildew-proofing fabrics.

TABLE 2.4
Organic fungicides

Chemical Group	Examples	Formula
Chlorinated phenols	TCP	OH, Cl, Cl, Cl
	PCP	OH, Cl, Cl, Cl, Cl, Cl
Nitrophenols	DNOC	OH, NO_2, CH_3, NO_2
Quinones	Chloranil	O, Cl, Cl, Cl, Cl, O
Chloronitrobenzenes	PCNB	Cl, Cl, Cl, Cl, Cl, NO_2
	TCNB	Cl, Cl, Cl, Cl, NO_2
Dithiocarbamates	Nabam	$NaS\overset{S}{\overset{\|}{C}}NH\,.\,CH_2\,.\,CH_2\,.\,NH\overset{S}{\overset{\|}{C}}SNa$
	Metham	$CH_3\,.\,NH\overset{S}{\overset{\|}{C}}SNa\,.\,2H_2O$

TABLE 2.4—*continued*

Chemical Group	Examples	Formula
Captan and analogues	Captan	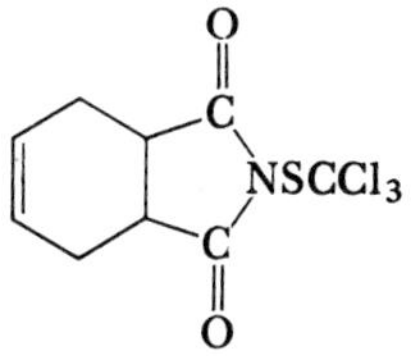 [cyclohexene ring fused to imide: C(=O)–N(SCCl$_3$)–C(=O)]

3.2.2. Nitrophenols. Nitrophenols are potent fungicides but are also toxic to higher plants and, as a result, find use as general herbicides. DNOC is toxic to mammals also.

3.2.3. Chloronitrobenzenes. The chloronitrobenzenes (e.g. PCNB) control post-harvest rots of fruits and vegetables. They are chiefly effective against soil fungi and act largely by retarding spore germination and colony growth. Incidentally, they are also useful for delaying sprouting of stored potatoes.

3.2.4. Quinones. The quinones, such as chloranil, are used to protect seeds from fungal attack. They generally have a low mammalian toxicity but, in sensitive people, may cause severe skin irritation.

3.2.5. Dithiocarbamates. The dithiocarbamate group of fungicides (thiram, nabam, vapam) is versatile, has a low animal toxicity and, as a result, is widely used.

3.2.6. Captan. Captan has found favour as a soil fungicide particularly for the control of *Pythium* species and has been added at levels of one percent to the diet of rats without apparent ill effects.

3.2.7. Antibiotics. Antibiotics have been considered for use as fungicides, but in most cases they are prohibitively expensive although there has been limited use of streptomycin, not as a fungicide but rather for the systemic control of plant bacterial pathogens.

3.3 Herbicides

The major groups of organic herbicides are shown in Table 2.5.

TABLE 2.5

Organic herbicides

Chemical Group	Examples	Formula
Chloroaliphatic acids	TCA	$Cl-C(Cl)(Cl)-COOH$
	Dalapon	$CH_3-C(Cl)(Cl)-COOH$
Chlorophenoxy- and chlorobenzoic acids	2,4-D	$Cl-C_6H_3(Cl)-O.CH_2.COOH$
	2,4,5-T	$Cl-C_6H_2(Cl)_2-O.CH_2.COOH$
	MCPA	$Cl-C_6H_3(CH_3)-O.CH_2.COOH$
	2,4-DB	$Cl-C_6H_3(Cl)-O.CH_2.CH_2.CH_2.COOH$
	2,3,6-TBA	$C_6H_2(Cl)_3-COOH$
Amides	CDAA	$Cl.CH_2.CO.N(CH_2.CH{=}CH_2)$
	Propanil	$Cl-C_6H_3(Cl)-NH.CO.C_2H_5$
Ureas	Monuron	$Cl-C_6H_4-NH.CO.N(CH_3)_2$

TABLE 2.5—*continued*

Chemical Group	Examples	Formula
	Fenuron	$NH.CO.N(CH_3)_2$
	Diuron	Cl, Cl; $NH.CO.N(CH_3)_2$
Carbamates	CIPC	Cl; $NH.CO.CH(CH_3)_2$
	Metham	$CH_3.NH.C(=S).S.Na$
	Di-allate	$(CH_3)_2CH$, $(CH_3)_2CH$ $>N.C(=O).SCH_2.C(Cl)=CH.Cl$
Triazines	Simazine	Cl; $C_2H_5.NH.C$ … $C.NH.C_2H_5$ (triazine ring: C, N, C, N, C, N)
Bipyridyliums	Diquat	N^+, ^+N; $CH_2—CH_2$; $2Br^-$
	Paraquat	CH_3N^+ … $^+NCH_3$
Toluidines	Dipropalin	CH_3; NO_2, NO_2; $N(C_3H_7)_2$
	Trifluralin	CF_3; NO_2, NO_2; $N(C_3H_7)_2$

TABLE 2.5—*continued*

Chemical Group	Examples	Formula
Substituted phenols	PCP	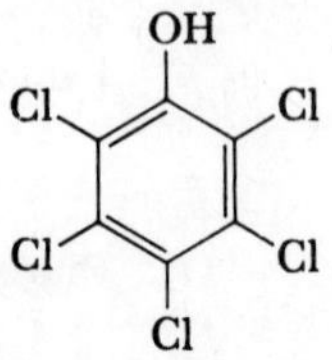 OH; Cl, Cl, Cl, Cl, Cl
	Dinoseb	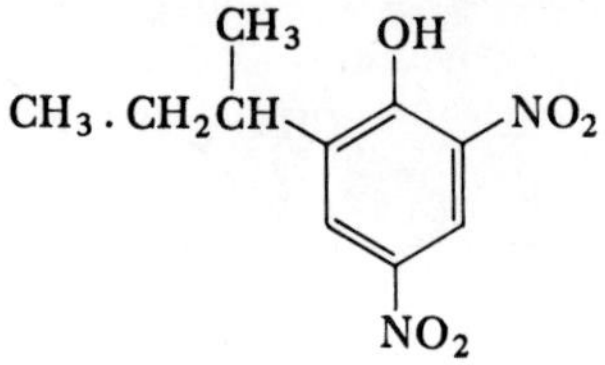 CH_3; $CH_3 . CH_2CH$; OH; NO_2; NO_2
Miscellaneous	Endothal	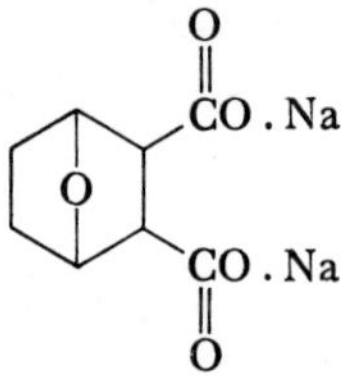 O; CO . Na; O; CO . Na; O
	NPA	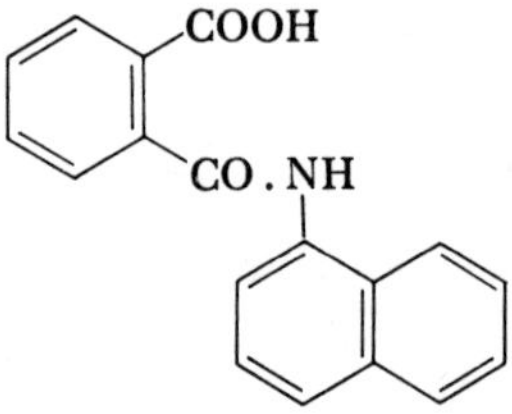 COOH; CO . NH
	Amitrole	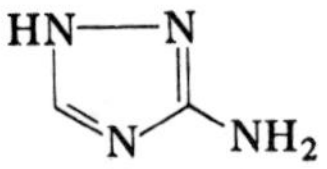 HN—N; N; NH_2
	Maleic hydrazide	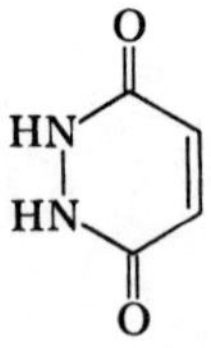 O; HN; HN; O
	Bromacil	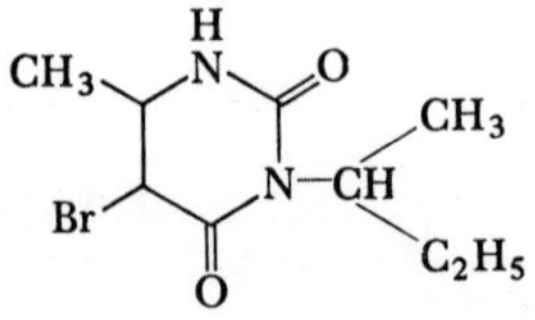 H; CH_3; N; O; CH_3; N—CH; Br; C_2H_5; O

3.3.1. Chloroaliphatic Acids. The sodium salts of chloroaliphatic acids (trichloracetic acid, dalapon) have been used for many years as soil treatments for controlling grasses. These compounds have a low toxicity for animals and most are rapidly degraded in the soil.

3.3.2. Chlorophenoxy and Chlorobenzoic Acids. It was the discovery of the auxin-like or growth regulatory properties of the chlorinated phenoxyacetic acids and their subsequent employment at herbicides that began the modern era of selective weed control. These compounds are cheap to produce, easily translocated in plants, selective to dicotyledons, and readily absorbed by roots from the soil. They destroy plants by causing gross disturbances in their development. A large variety of derivatives has been produced which fulfills a range of requirements for weed control and those which are extensively used today include 2,4-D, 2,4,5-T and MCPA. Another, 2,4-DB, is an interesting example of a carefully designed, highly selective herbicide which in its original molecular state it is only slightly toxic to plants. However, many weeds will rapidly oxidize it to the herbicidally-active compound 2,4-D. Other plants, especially legumes, either convert it very slowly or not at all. As a result 2,4-DB is used to control weeds in leguminous crops such as pea, clover, trefoil and lucerne.

The chlorobenzoic acids (2,4,6-TBA, dinoben) are much more persistent in the soil than the chlorophenoxy acids and are therefore very useful for controlling deep-rooted perennial weeds.

3.3.3. Amides. A variety of simple aliphatic amides show herbicidal activity. Most of the more recently developed compounds are anilides (derivatives of acetamide, CH_3CONH_2) such as propanil. They all function by inhibiting photosynthesis and have low mammalian toxicity.

3.3.4. Ureas. Some simple ureas have herbicidal activity but the agriculturally important ones are the phenyl derivatives such as monuron and its analogues. Some of the ureas are very persistent, indeed monuron and diuron are amongst the most recalcitrant of herbicides and may remain effective for several years. They must therefore be used with great care to avoid rendering soils unusable for long periods.

3.3.5. Carbamates. The carbamate herbicides include a number of esters of carbamic (NH_2COOH) and thiocarbamic acid (NH_2CSOH).

They all act by inhibiting cell growth and photosynthesis and have low mammalian toxicities.

3.3.6. Triazines. Although simazine was synthesized as long ago as 1885, the triazine herbicides were not introduced until 1955. These compounds inhibit photosynthesis and function at low dose rates—exposure of roots to 0.25–1.0 ppm stops all photosynthetic activity. They are particularly useful as pre-emergent herbicides, and many of them are extremely persistent in the soil. The persistence of these compounds together with the fact that they do not seem to have any adverse effects on non-plant life makes them attractive for long term protection.

3.3.7. Bipyridyls. It was the discovery that quaternary ammonium germicides, such as cetyltrimethylammonium bromide, desiccated young plants that led to the development of the bipyridilium herbicides. The two best known are paraquat and diquat and these are throught to be reduced to free radicals in the plant which, upon reoxidation, form highly toxic peroxide radicals. These herbicides are functionally very short lived after application, largely because they are highly cationic and are rapidly immobilized by the soil particles (see page 51).

3.3.8. Toluidines. The first commercially important member of the toluidines was dipropalin, with its analogue trifluralin as a more recent development. Both are effective herbicides with low mammalian toxicities and are extremely persistent. These characteristics make them, like the triazines, suitable for long-term protection.

3.3.9. Miscellaneous. Some substituted phenols, such as PCP and dinoseb, are used as contact pre-emergent herbicides in the control of annual weeds.

Endothal is a defoliant and desiccant and the phthallic acid derivative, NPA, a pre-emergent herbicide.

Amitrole, which is readily absorbed through the leaves, inhibits photosynthesis and is very effective for direct foliar application to eradicate poison ivy, poison oak, scrub oaks and thistles.

Maleic hydrazide has growth regulatory properties. It inhibits mitosis and respiration and is used mainly to prevent the development of suckers on tobacco and cranberries and to retard the growth of trees, shrubbery and grass.

Derivatives of the pyrimidine, uracil (e.g. bromacil) are selective for many annual weeds.

3.4. Other Pesticides
Some insecticides, particularly carbamates and organophosphates, are effective molluscicides. Metaldehyde [$(OCH{-}CH_3)_4$] is a specific attractant and toxicant for garden snails and slugs.

Plant parasitic nematodes attack the roots of crop plants and cause vast amounts of damage. Many nematocides used to control them are in fact soil fumigants, particularly, the halogenated hydrocarbons such as dichloropropene, dichloropropane, 1,2-dibromoethane and methyl bromide. Further examples of nematocides include the carbamate, vapam (sodium *N*-methyl dithiocarbamate, ($CH_3NH(S)SNa$) and more recently some organo-phosphorus compounds, such as zinophos.

Finally, important rodenticides include "red squill", which contains cardiac glycosides causing convulsions and respiratory failure; strychnine, (a complex dibasic alkaloid); sodium fluoracetate (FCH_2COONa), and hydroxycoumarin analogues, such as the anticoagulant warfarin [III].

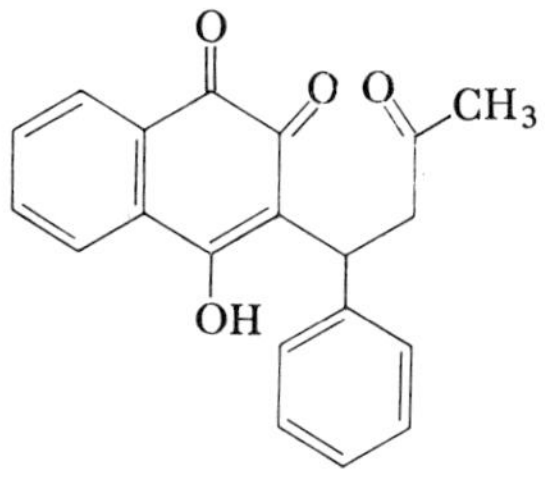

[III]

ANALYTICAL METHODS

For the safe and effective utilization of pesticides it is important that there are sensitive methods for determining their levels in the environment. There are two major categories of analyses; the first being quality control on the part of the manufacturer or governmental laboratory; the second detection and quantification of pesticides in the various segments of the biosphere. The actual methods involved are essentially similar in both cases, excepting that in environmental work more sophisticated extraction and purification methods are required to isolate the specific chemical from the vast array of other substances found, for example, in soil or biological tissue. Before the organic pesticide "revolution" the pesticide chemist was called upon to perform analyses for a limited number of residues such as lead, fluoride, pyrethins and rotenone. With the introduction of synthetic organic pesticides he was asked to measure a rapidly growing plethora of complex substances and it is only in recent years that the technology of chemical analysis has begun to catch up with the imagination of the organic chemist. It is primarily due to advances in instrumentation that the modern analytical chemist has at his command an impressive array of selective and sensitive tools, originally developed to study fundamental physico-chemical properties rather than for pesticide analysis.

1. Sampling of Contaminated Material

Sampling is of prime importance in residue studies and clearly any analysis, however accurate, only gives meaningful data if the sampling is representative. Therefore statistical considerations are important when deciding on the number, type, and location of samples.

2. Extraction of Pesticide from Sample

As a rule the pesticide must be extracted with organic solvents from complex heterogenous systems, such as soil, plant or animal tissue. Before any subsequent assessment of residue levels can be made it is necessary to know the efficiency of the particular extraction procedure; an efficiency that is dependent upon the methods used and the materials involved. For example, with soil samples, a proportion of the pesticide may be unavailable for assay because of strong adsorption on or within soil particles. The recovery rate can be determined by adding known amounts of the pesticide in question to known amounts of sample and running a standard extraction procedure. Non-amended controls are necessary to adjust for endogenous levels.

Where biological materials are involved, extraction procedures must be carefully designed so that only those residues located in the specific site under investigation are removed. For example, with plant tissue it may be important to discriminate between pesticides deposited on the outer surfaces of leaves and those incorporated into leaf tissues. In these circumstances a preliminary "mild" process is necessary to remove the former material followed by maceration in an organic solvent to extract the latter.

Before analysis it is often prudent to "clean-up" the extract which, if of plant or animal origin, may contain large amounts of impurities (lipids in particular). The pesticides can usually be separated either by partitioning between polar and non-polar solvents, or by thin-layer or column chromatography. Pesticide extracts from soil contain quantities of coloured humic material which must be removed before successful analysis can be achieved.

Having obtained a "clean" extract, how does one identify and quantify the various pesticide components? There are three basic types of measurements available to the analyst—biological, chemical and physical. Biological methods are obligatory for efficacy and toxicity tests but for quantitative assay they are generally both time consuming and inaccurate. They are therefore being rapidly superseded by chemical and physical methods.

3. Biological Assay

Biological assay (bioassay) has traditionally been the most important technique used in the estimation of herbicides and insecticides. It

A. Dose/response Curve.

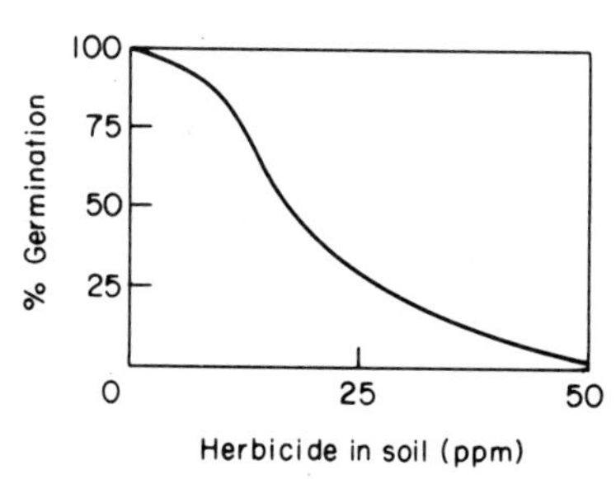

B. Experimental results.

	seeds planted at day:					
	1	7	14	21	28	35
% germination	1.2	4.3	4.7	29.6	76.5	96.8

C. Herbicide persistence.

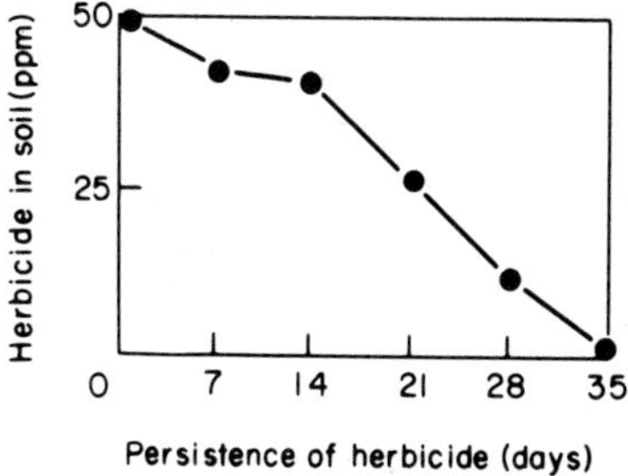

Fig. 2.2. A bioassay experiment.

depends on there being a predictable relationship between dose and biological response when the indicator material and environmental conditions are constant. For example, a common approach to herbicide evaluation is to measure promotion or inhibition of root and/or shoot segment elongation in the presence of various concentrations of the chemical. Intact plants rather than tissue can also be used. Once a dose/response relationship is established the concentration of herbicide in extracts from plant, animal, water or soil may be assessed. One of the main difficulties in using this type of bioassay is the great weight put upon successful extraction and purification procedures. For example, the materials from which the chemical is extracted may also contain endogenous substances or toxic breakdown products of the parent compound both of which may induce an inappropriate response in the indicator. It is therefore necessary to be aware of these drawbacks, to obtain "clean" extracts and to run appropriate controls if reliable results are to be achieved.

A more direct bioassay method, which allows for the determination of pesticide levels *in situ*, is to measure the response of a plant or animal exposed to a contaminated sample. For instance, using this technique a characteristic (such as rate of and percent germination, plant dry weight, number of leaves, legume nodulation, insect reproduction rate, etc.) is measured against known levels of pesticide in a soil sample. Then, by exposing the indicator to pesticide-amended soil, say at weekly intervals, it is possible to follow the efficacy of that compound over a period of time (Fig. 2.2). It should be remembered that this second type of bioassay only measures the availability of a pesticide to the selected indicator. This is not necessarily synonymous with the absolute level of pesticide remaining in the sample as the physico-chemical characteristics of the soil have an important bearing on pesticide availability.

4. Chromatography

Chromatography is an important technique in residue analysis, both for the separation of mixtures of pesticides and for their identification and quantification. The most widely used of these methods are thin-layer chromatography (TLC) and gas-liquid chromatography (GLC). In addition, high pressure liquid-chromatography is beginning to be applied to residue problems; it is rapid and very sensitive yet expensive. The use of the gas-liquid chromatograph has expanded greatly in recent years and is now a standard technique in biological and chemical laboratories. It is an extremely versatile instrument, in that it can be used to quantify any substance that is both sufficiently volatile (or can be made so by derivative formation) and stable as a vapour. In addition GLC analysis is very rapid (elution times of a few minutes), sensitive (one nanogram or less) and provides both qualitative and quantitative data simultaneously. However, it is also a technique that is easily misused by the inexperienced and it has been said that, during the last two decades, more misinformation has resulted from inadequately performed gas chromatography than from any other single technique used in residue analysis. Once again carefully controlled sample preparation is essential for reproducible results. Erroneous results can be caused by interfering substances in the extracts although those analyzed by electron capture and thermionic detectors frequently require little purification. Since the method is so sensitive, it is of paramount importance to use scrupulously clean apparatus (columns, inlet ports, syringes, detectors, etc.).

5. Spectrophotometry

A variety of spectrophotometric methods are used for assay and identification of pesticides and their products. Some compounds give a specific colour on reacting with other substances and this can be measured in the visible region (380–780 nm). Others absorb in the ultra-violet region (185–380 nm). The intensity of absorption is directly proportional to the concentration of the pesticide. Ultraviolet (UV) spectroscopy has some uses in identification, whilst infra-red (IR) spectroscopy (1-50 μm) is of great value. Until recently it was necessary to have relatively large amounts of sample (at least 100 μg) for IR work but this problem has been overcome, to some extent, by using microcells together with beam-condensing systems which reduce the size of sample necessary to obtain spectra. It is even possible to obtain IR spectra directly from gas chromatograph effluents.

Perhaps the most promising development in recent years for residue analysis is the combined gas chromatograph-mass spectrometer (GC-MS). This marriage of techniques allows the analyst to separate complex mixtures of pesticides and breakdown products and to measure and identify each component in one operation. The drawbacks with GC-MS are the high cost of the equipment and the expertise required to operate the machinery. The mass spectrometer is the most sensitive of all spectroscopic tools and it is often possible to obtain a spectrum on as little as ten nanograms of sample.

Provided that a reference spectrum is available an unknown residue can be identified immediately on the basis of the uniqueness of its mass spectrum. In spite of the potency of this method it is only very recently that it has begun to make a significant contribution to pesticide research but its value will no doubt increase greatly during the next few years.

6. Enzymic Methods

Enzymic methods are frequently employed for measuring pesticide residues, a notable example being the use of acetylcholine esterase in insecticide assays. Many insecticides function by inhibiting acetylcholine esterase, an enzyme essential for normal neurotransmission. It is possible, therefore, to devise assays for these compounds using purified acetylcholine esterase and measuring the amount of enzyme inhibition.

7. Miscellaneous

Labelled pesticides and their breakdown products can be detected and measured rapidly and, as a result, radioactive tracers are commonly used to elucidate the fate of pesticides after they have been applied to the environment. Experiments with radioactive pesticides may be designed to (a) relate the rate of pesticide metabolism or degradation to the duration of its functional activity, (b) investigate the rate of elimination from the mammalian body and (c) evaluate the rate of metabolism, sites of accumulation and the nature of the metabolic products in plants, animals and soils.

The chlorinated hydrocarbon and organophosphorus pesticides are sometimes determined by measuring the total organic chlorine or phosphorus respectively. Although relatively simple, these methods are not specific for individual compounds but have found applications in routine analyses in the field.

PESTICIDE DECAY

1. Non-Biological

1.1. Photodecomposition

Most of the pesticides which absorb visible or ultra-violet radiation undergo some degree of decay when exposed to light. This characteristic has been demonstrated in laboratory experiments for monuron, simazine, amiben, DDT, dieldrin, trifluralin, paraquat and others. However, when considering the fate of pesticides in the environment, photodecomposition is probably only significant for a few of the chemically less stable pesticides. It is also obvious that light-stimulated transformations can only occur where light can penetrate. Therefore photodecomposition in soil, for instance, can only be important at the surface and even then may be difficult to distinguish from all the other factors contributing to the decay of the pesticide.

In many experiments only the radiation-induced disappearance of the original chemical has been monitored, whereas in other cases, the products of the reaction have been identified. For example paraquat is photodecomposed to methylamine hydrochloride (Fig. 2.3), and 2,4-D destruction eventually leads to the formation of humic acid, a normal polymeric constituent of soil (Fig. 2.4).

$$\left[H_3C-N\langle\bigcirc\rangle-\langle\bigcirc\rangle N-CH_3\right]^{2+} 2Cl^- \longrightarrow \left[H_3C-N\langle\bigcirc\rangle-COOH\right]^{+} Cl^-$$

$$\downarrow$$

$$CH_3NH_2 \cdot HCl$$

Fig. 2.3. Photodecomposition of paraquat.

It is probable that radiation-induced reactions facilitate the decay of some pesticides by making them more vulnerable to microbial attack. In addition, the sensitivity of a variety of compounds to light may be enhanced by their association with dyes, pigments, zinc oxides, ferric and uranyl compounds and cobalt complexes.

1.2. Catalysis

The catalytic decay of pesticides by non-living organic and inorganic components of the environment is well documented. For example, in one type of habitat soil, the mineral and organic fractions catalyse a variety of pesticide transformations involving reduction, oxidation, hydrolysis and isomerization. Under conditions of poor aeration humic material may stimulate the reduction of some functional groups in pesticides. Basic amino acids and reduced iron porphyrins may catalyze the hydrolysis of organophosphorous insecticides and the dehydrochlorination of DDT.

Non-biological oxidation, mediated by the inorganic soil fraction, is observed during the conversion of aldrin to dieldrin. Carbon-

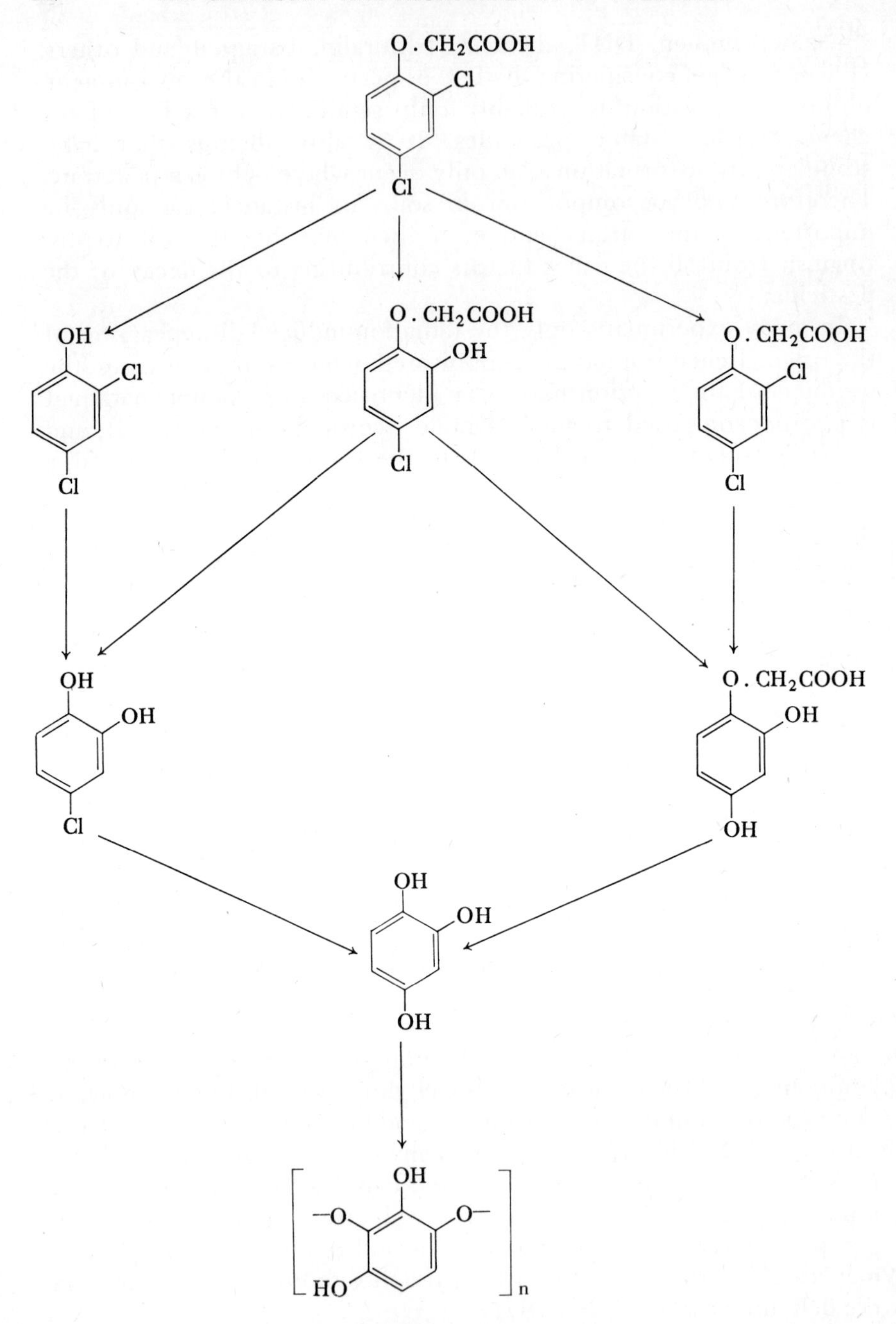

Fig. 2.4. Photodecomposition of 2,4-D.

ates and sulphides of iron, manganese and cobalt are capable of catalyzing oxidation and reduction reactions, whilst cupric ions are responsible for the hydrolysis of some organophosphorus esters. In addition, acidic and basic sites on clay particles can catalyse chemical reactions as demonstrated by the isomerization of endrin and the hydrolysis of atrazine and DDT.

2. Microbiological

2.1. Persistence and the Role of Microbes

It is now clear that the disappearance of many organic pesticides from the soil is largely due to the activities of microorganisms. On what evidence does this statement rest? When the persistence of pesticides in the soil was first studied it was readily apparent that those conditions which favour bacterial growth often coincide with the most rapid rates of pesticide disappearance and that there are considerable differences in decay rates between soil types related to biological (and non-biological) variations.

Moreover, there is often a relationship between the actual numbers of bacteria in the soil and the rate of disappearance of a pesticide, whilst more direct evidence of the microbial detoxication of pesticides comes from experiments in which the inhibition of microbial activity results in much reduced rates of degradation. This inhibition may be achieved in two ways, firstly by autoclaving soil, although this also affects the soil physical and chemical characteristics; and secondly by the addition of microbial inhibitors. For example, small concentrations of the cytochrome oxidase inhibitor, sodium azide and of the respiratory inhibitor, sodium fluoride can completely prevent the breakdown of 2,4-D,MCPA and 4-CPA in enriched soils.

For a long time the final proof of the dominant role of soil bacteria has been regarded as the isolation of the responsible organism, its growth in pure culture using the pesticide as the sole source of carbon or nitrogen, and the demonstration that it can inactivate the pesticide perfused through a column of soil containing the relevent microbe. This is perhaps less than conclusive, because mixed cultures of microbes may degrade a particular compound whereas no single species is able to. Also, in recent years, it has become evident that pesticides can be altered or even completely degraded by co-oxidation (co-metabolism) during which microbes effect transformations without being able to grow on the pesticides and often without deriving carbon or energy from them.

TABLE 2.6

Some microbial genera implicated in the degradation of herbicides

Pseudomonas	*Corynebacterium*	*Achromobacter*	*Bacillus*	*Arthrobacter*	*Flavobacterium*
TCA	2,4-D	2,4-D	Dalapon	TCA	Dalapon
Dalapon	Paraquat	2,4,5-T	Monuron	Dalapon	2,4-D
2,4-D		MCPA		2,4-D	2,4-DB
Monuron				2,4,5-T	Maleic Hydrazide
PCP				Propanil	
Paraquat				DNOC	
Diquat					
DNOC					

Nocardia	*Trichoderma*	*Aspergillus*
Dalapon	TCA	Dalapon
2,4-D	Dalapon	Monuron
2,4-DB	Picloram	Trifluralin
Propanil		Picloram
		2,4-D

An early demonstration that bacteria can metabolize organic pesticides occurred in 1951 when it was shown that a strain of *Arthrobacter globiformis* grows on agar containing 2,4-D. Since that time there have been many similar examples. Table 2.6 presents just some of the microorganisms that have been shown to grow on pesticides as sole carbon, energy and sometimes nitrogen sources. It can be seen that the capacity to do this is widespread amongst microorganisms, species of *Pseudomonas*, *Nocardia* and *Aspergillus* being the most adept. Indeed it is remarkable that so many organisms can metabolize not only chlorinated organic compounds, since there are few of these in Nature, but also all sorts of novel substances with which they have had no previous contact. This faculty may arise in two ways; firstly by induction of enzymes that the organism already has the genetic competence to produce and secondly by mutation leading to altered enzymes or their control mechanisms.

Although microbes do have these extraordinary adaptive capacities, it seems that they are not infallible and some pesticides are degraded only very slowly or appear completely immune to microbial attack. The most commonly described of these recalcitrant molecules is DDT which is extremely persistent, especially in aerobic environments, and indeed nobody has yet found a microbe that can grow on it as a sole carbon and energy source.

2.2. Fundamental Mechanisms of Decay

Microbes catalyse a number of basic chemical alterations to pesticide molecules. These are summarized below and specific examples can be found in later sections.

2.2.1. Dehalogenation. Dehalogenation is frequently involved in pesticide breakdown. It is the first reaction in the biodegradation of several aliphatic acids and is implicated in the decomposition of some of the chlorinated hydrocarbon insecticides. Enzymes that catalyse these reactions have been isolated from various microorganisms.

2.2.2. Dealkylation. Dealkylation occurs to several pesticides that have alkyl groups attached to nitrogen, oxygen or sulphur atoms (for example, triazines and toluidines). Alkyl groups attached to carbon atoms are, in general, resistant to microbial attack.

2.2.3. Amide and Ester Hydrolysis. Many pesticides are esters of inorganic acids—such as the phosphate ester insecticides (parathion, malathion)—or are amides like the phenylamine herbicides (phenylureas, phenylcarbamates, acylanilides). Several microbes have been shown to hydrolyse the amide and ester bonds in these compounds.

2.2.4. Oxidation. Many microorganisms synthesize oxygenase enzymes which will introduce molecular oxygen into organic molecules, particularly those with aromatic rings. It is common for pesticides to be oxidized in this manner and it involves the insertion of an hydroxyl group or the formation of an epoxide.

2.2.5. Reduction. The main reductive transformation mediated by microbial activity involves the reduction of nitro (NO_2) groups to amine (NH_2) groups. In addition, quinones are reduced to phenols by microbial thiol compounds.

2.2.6. Ring Cleavage. Aromatic rings are degraded by many soil bacteria and fungi. Initially the rings are hydroxylated by monooxygenase enzymes (as in 2.2.4.) forming catechols (aromatic compounds with two hydroxyl groups in adjacent positions) and are then cleaved by dioxygenase enzymes to yield muconic acids or muconic semialdehydes.

2.2.7. Condensation or Conjugate Formation. Condensation involves the coupling of the toxic molecule, or a portion of it, to another organic .compound, often with the result that the pesticide or derivative is inactivated.

2.3. Co-Metabolism

Many detoxication reactions occur whilst the microbe is utilizing the pesticide as its source of carbon and energy. But microbes can also bring about chemical alterations of pesticides without deriving sufficient carbon or energy for growth from these reactions. This fortuitous process has been called co-metabolism or co-oxidation, and is thought to reflect a lack of substrate specificity of some of the microbial transport mechanisms and enzymes. Clearly for co-metabolism to occur the microbes must obtain the bulk or all of their carbon and energy from other substrates.

The phenomenon of co-metabolism, in relation to environmental pollution, is important because it presents the possibility that a series of co-metabolic reactions involving several microorganisms can lead to the total degradation of a pesticide. A pesticide which, in itself, is unable to support microbial growth. Compounds such as DDT have been described as recalcitrant, partly because it has not been possible to isolate any microbe capable of utilizing them as sole carbon and energy sources. There is evidence however that suggests that two micro-organisms (*Aerobacter aerogenes* and an *Hydrogenomonas* species) can *together* convert DDT, in liquid culture, to *p*-chlorophenylacetic acid by co-metabolism (Fig. 2.5) although it is not yet known whether this occurs in the environment.

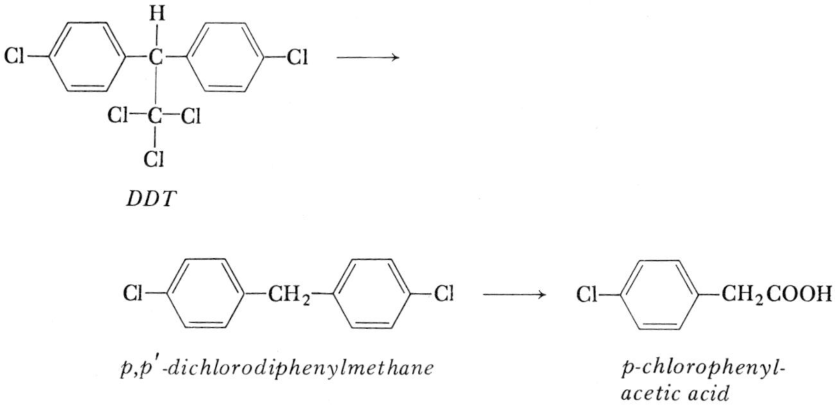

DDT

p,p'-dichlorodiphenylmethane → *p-chlorophenyl-acetic acid*

Fig. 2.5. Co-metabolism of DDT.

A further example concerns the herbicide, 2,3,6-TBA and the closely related 2,4,5-T which can both be co-metabolized to 3,5-dichlorocatechol. The last named compound is an intermediate in the breakdown of 2,4-D by an *Arthrobacter* species and can also be co-metabolized by an *Achromobacter* species. This means that 2,3,6-TBA and 2,4,5-T may be degraded by a combination of metabolism and co-metabolism (Fig. 2.6). Although it is easy to demonstrate, *in vitro*, the likely importance of co-metabolism in pesticide degradation, it is difficult to obtain direct evidence for its operation in the environment. For instance, in a series of experiments it was shown that microbes in lake water, albeit in the laboratory, degraded 2,3,6-TBA without any increase in their population. In addition, if benzoic acid was added (to enrich for aromatic ring-degraders) there was an acceleration in the rate of 2,3,6-TBA degradation although there were still no organisms present capable of growing on it.

In conclusion, it is likely that there will be many reports of the involvement of co-metabolism in pesticide degradation in future years and the lesson the mechanism teaches is that it is not an essential criterion of a pesticide's biodegradability that it should be capable of serving as a sole carbon and energy source for microbial growth.

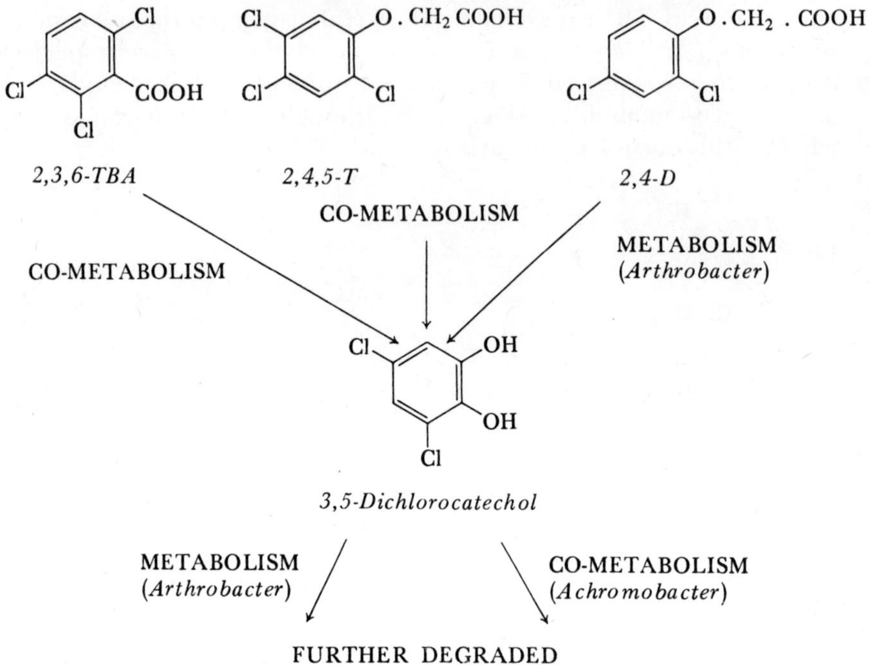

2,3,6-TBA — CO-METABOLISM →

2,4,5-T — CO-METABOLISM →

2,4-D — METABOLISM (*Arthrobacter*) →

3,5-Dichlorocatechol

METABOLISM (*Arthrobacter*) → FURTHER DEGRADED

CO-METABOLISM (*Achromobacter*) → FURTHER DEGRADED

Fig. 2.6. Microbial degradation of 2,3,6-TBA, 2,4,5-T and 2,4-D.

2.4. Insecticide Degradation

There is relatively little information about the microbial metabolism of insecticides as compared to that concerning their metabolism in animals and plants.

2.4.1. Chlorinated Hydrocarbons. Since this group of compounds is rather persistent in the environment their biodegradation has attracted a relatively large proportion of research effort.

The most extensively studied aspect of DDT metabolism has been microbial dechlorination, which involves loss of one of the chlorine atoms from the trichloro-methyl group to form TDE. This is a function of many microorganisms. The additional possibility of DDT degradation by co-metabolism has been mentioned above. Little information is available on the metabolism of DDT analogues, although about twenty products of its incomplete breakdown have been isolated from either laboratory cultures or the natural environment.

Benzene hexachloride (BHC) is metabolized slowly by bacteria, one of the products being benzene itself.

The chlorinated cyclodienes are persistent but nevertheless appear to be slowly degraded by soil microbes. Many microbes can form epoxides and so convert aldrin to dieldrin which can then be further degraded although the components of the metabolic pathway are unknown. Little detail is available about the routes of degradation of endrin, heptachlor and chlordane. However, heptachlor is converted to its epoxide by soil microbes (Fig. 2.7).

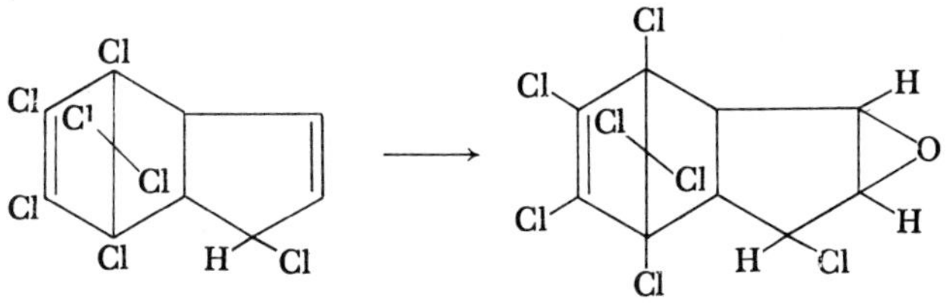

Heptachlor → *Heptachlor epoxide*

Fig. 2.7. Microbial oxidation of heptachlor.

2.4.2. Organophosphates. Detailed studies of organophosphorus insecticide degradation are limited and, although most of this group are known to degrade quite rapidly in soil, the relative importance of the microflora in this process is not clear. The initial step usually involves an esterase enzyme, and dialkylphenylphosphates together with thiophosphates are degraded by various microorganisms, including *Pseudomonas fluorescens* and *Thiobacillus thiooxidans*. The breakdown of phorate is summarized in Fig. 2.8.

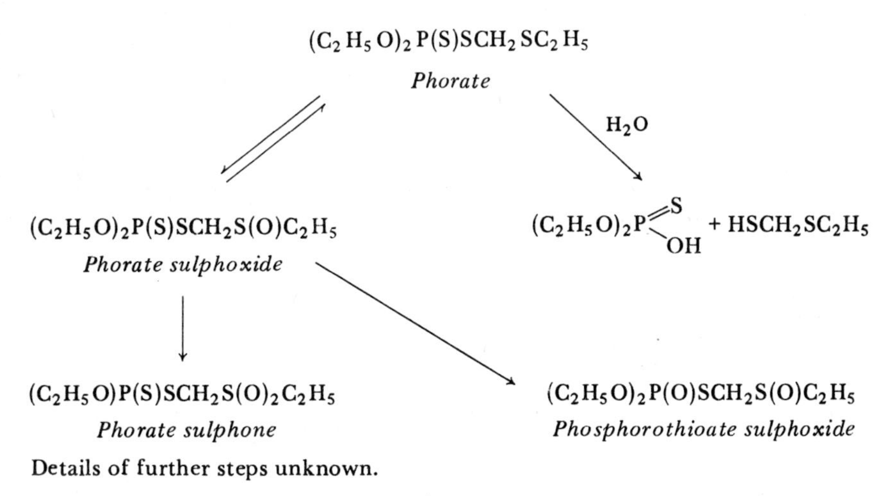

Fig. 2.8. Phorate breakdown by soil bacteria.

Pseudomonas melophthora can synthesize an esterase which allows it to degrade parathion and diazinon whilst species of *Arthrobacter* and *Streptomyces* have also been shown to degrade diazinin. *Trichoderma viride* and a *Pseudomonas* species are known to catabolize malathion (Fig. 2.9).

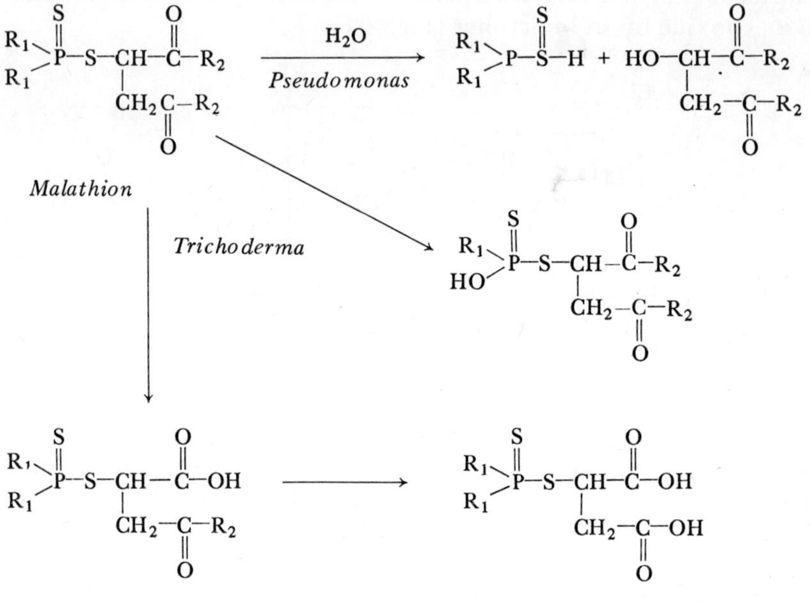

H_2O

Pseudomonas

Malathion

Trichoderma

Details of further steps unknown.

Fig. 2.9. Malathion breakdown by soil bacteria.

2.4.3. Carbamates. Microorganisms are involved in the rapid degradation of carbamates in soil. There are a number of species implicated in the breakdown of the most widely used of this group carbaryl, a derivative of naphthalene. The main catabolic routes for the degradation of both carbaryl and naphthalene converge at 1,2-dihydroxynaphthalene and are shown in Fig. 2.10.

2.5. Fungicide Degradation

2.5.1. Quinones. The quinone fungicides, such as chloranil, are rapidly degraded by microbes. They react very readily with thiol compounds and, in this way, are reduced to their corresponding phenols.

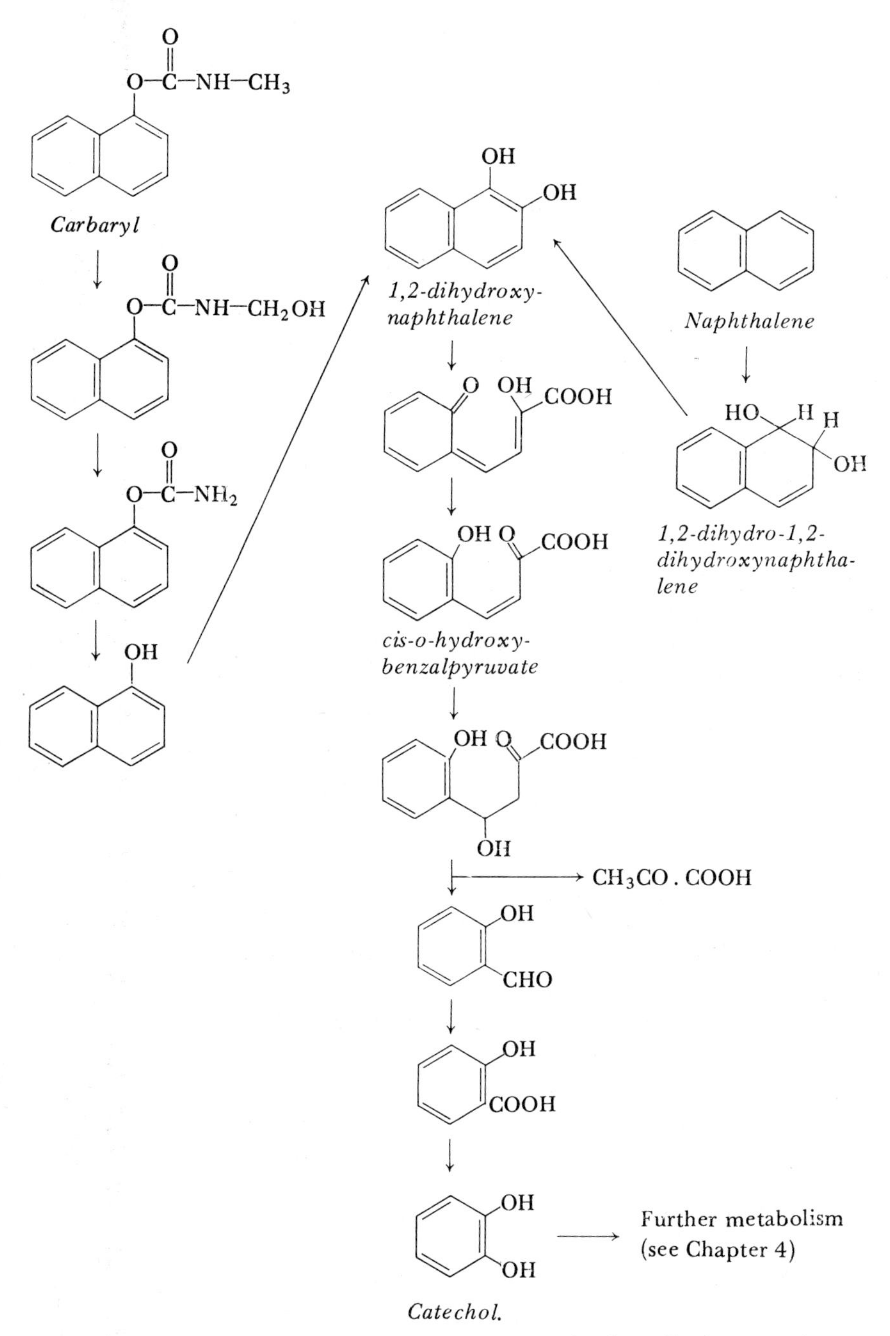

Fig. 2.10. Carbaryl and naphthalene degradation by soil microorganisms.

2.5.2. Organomercurials. Several organomercurials have been shown to be degraded by species of *Penicillium, Aspergillus* and *Trichoderma*. However, there is no detailed information concerning the pathways used.

2.5.3. Dithiocarbamates. Although there is some evidence for the microbial degradation of the dithiocarbamate, thiram, the decay of nabam in the soil is independent of microorganisms since the decay curves are similar with both fresh and sterilized soils.

2.5.4. Captan. Captan is rapidly hydrolyzed in the soil but there is no evidence for microbial involvement.

2.5.5. Phenols and Related Compounds. The ability to degrade both chlorinated and nitro-phenols is found in a variety of bacteria and fungi. The number of substituents and their position in the ring have a great effect on their biodegradability. In the case of halogen substituents, the highly halogenated phenols are the more stable and substitution in the *meta* position reduces the rate of microbial attack. There seem to be two ways in which microbes deal with the nitro-substituent in nitro-phenols. *Escherichia coli*, for example,

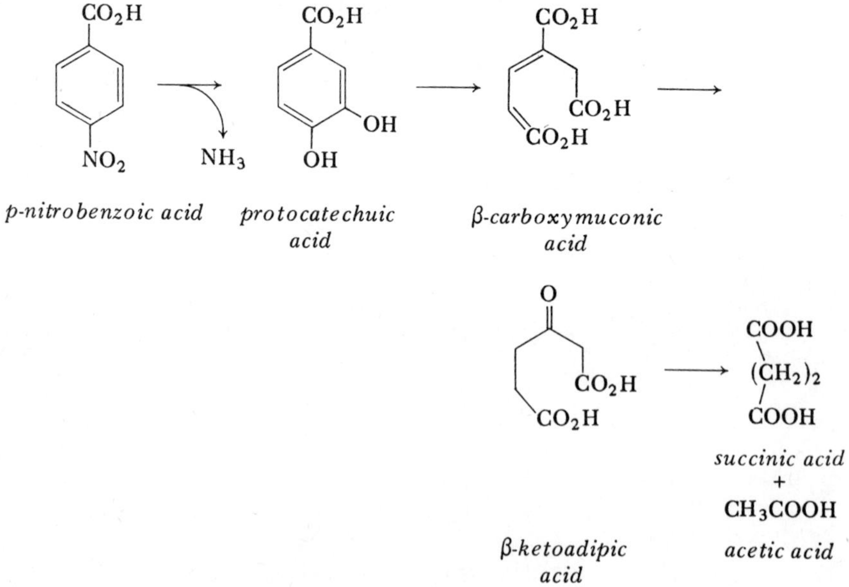

CO_2H NO_2 → NH_3 ; CO_2H OH OH → CO_2H CO_2H CO_2H →

p-nitrobenzoic acid *protocatechuic acid* *β-carboxymuconic acid*

O CO_2H CO_2H → COOH $(CH_2)_2$ COOH

succinic acid
+
CH_3COOH

β-ketoadipic acid *acetic acid*

Fig. 2.11. Degradation of *p*-nitrobenzoic acid by a *Nocardia* species.

reduces the nitro-group to an amino group whilst some *Nocardia* species remove the nitro-group, the nitrogen probably being converted to ammonia. For example, one species of *Nocardia* has been shown to degrade *p*-nitrobenzoic acid in this manner (Fig. 2.11). A strain of *Corynebacterium simplex* will use DNOC as its sole carbon and energy source, but the degradative route has yet to be determined.

2.5.6. Chloronitrobenzenes. There is little evidence that microbes degrade chloronitrobenzenes although a soil bacterium has been isolated that grows on the analogue, 1-chloronaphthalene which is subsequently degraded via 3-chlorosalicylic acid (Fig. 2.12). PCNB is

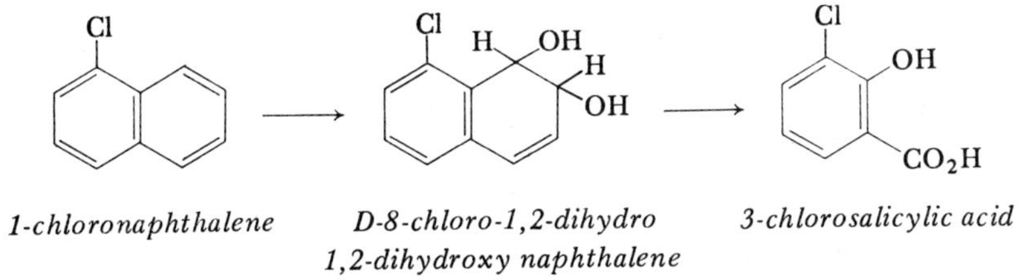

Cl → Cl H OH H OH → Cl OH CO_2H

1-chloronaphthalene *D-8-chloro-1,2-dihydro 1,2-dihydroxy naphthalene* *3-chlorosalicylic acid*

Fig. 2.12. Degradation of 1-chloronaphthalene.

reduced by a variety of microbes to pentachloroaniline (PCA). Recently it has been suggested that other species may catalyze the reverse process (Fig. 2.13).

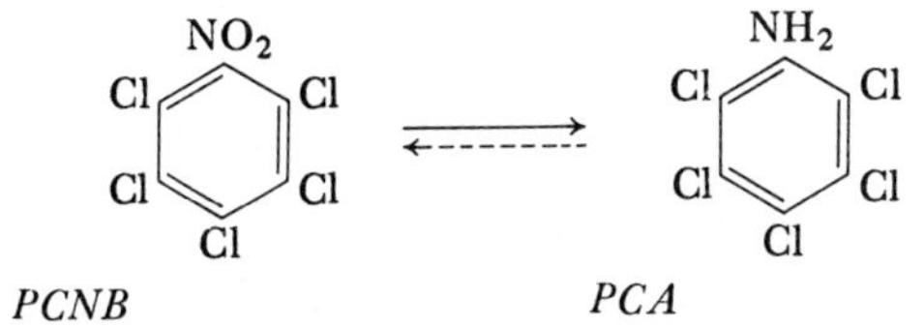

NO_2 Cl Cl Cl Cl Cl ⇄ NH_2 Cl Cl Cl Cl Cl

PCNB *PCA*

Fig. 2.13. Microbial reduction of PCNB.

2.6. Herbicide Degradation

In general there is more information concerning the microbial degradation of herbicides and their rates of decay (Fig. 2.14) than any other group of pesticides.

2.6.1. Chlorophenoxy Acids. 2,4-D is rapidly degraded in the soil, MCPA and 2,4,5-T more slowly. At least sixteen species of bacteria are known to degrade totally phenoxyacetic acids whilst the fungus

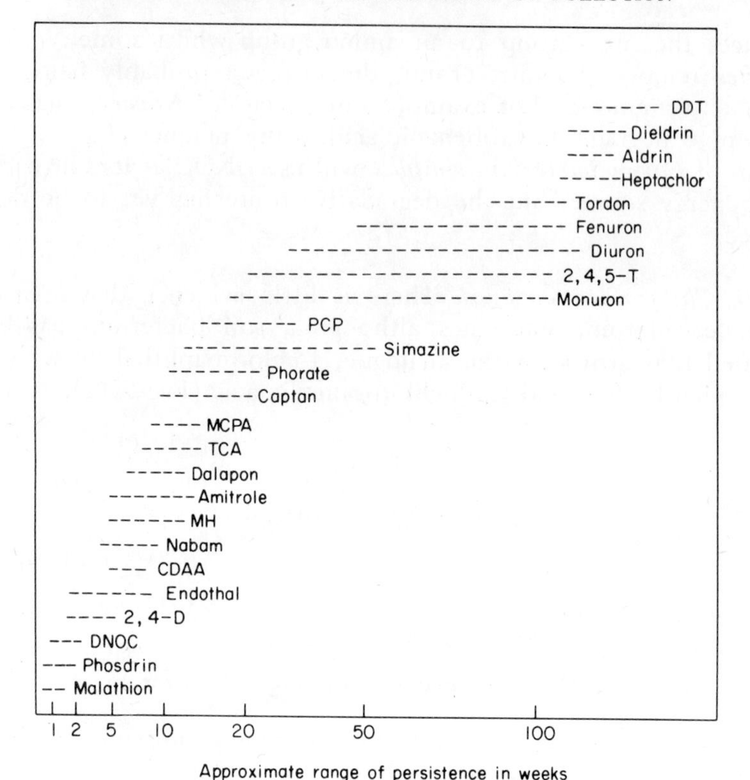

Fig. 2.14. Persistence of pesticides in soil.

Aspergillus niger attacks these compounds but only as far as hydroxylating the aromatic ring.

The synthesis of enzymes that degrade one of these compounds often confers upon the microbe the ability to degrade other phenoxyalkanoic acids. This is a common phenomenon when microbes are presented with close analogues of the inducing substrate. There seem to be two main pathways for the microbial degradation of phenoxyacetic acids; namely degradation via an hydroxyphenoxyacetic acid intermediate and degradation via the corresponding phenol. Figs. 2.15 and 16 show examples of these, the degradation of MCPA by an unidentified soil bacterium and the degradation of 2,4-D by an *Achromobacter* strain.

The biodegradability of the phenoxyacetic acids is dependent upon their chemical structure (Fig. 2.6) and is determined by the following criteria:

A. Chlorine substitution in the *para* position which makes the compound more labile.

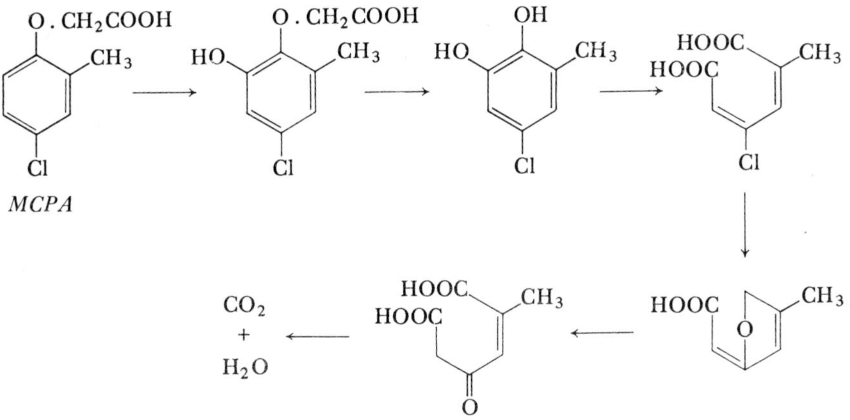

Fig. 2.15. Microbial degradation of MCPA by an unidentified microbial species.

B. Chlorine substitution in the *ortho* position which leads to a very recalcitrant molecule.

C. *Meta* substitution which also decreases biodegradability, but less so than *ortho* substitution.

D. *Para* substitution, in di- and tri-substituted compounds, which overcomes the deactivating effect of other substituents.

E. Methyl substitution which decreases biodegradability more than chlorine substitution.

The higher phenoxyalkanoic acids are metabolized in a similar way to the phenoxyacetic acids after β-oxidation of the alkanoic side chain. The rate of side chain degradation is also reduced by *meta* substitution of the ring. An unidentified *Flavobacterium* species is known to cleave the ether link in 2,4-DB as the first step in its attack on the molecule, thereby completely inactivating the herbicide (Fig. 2.17).

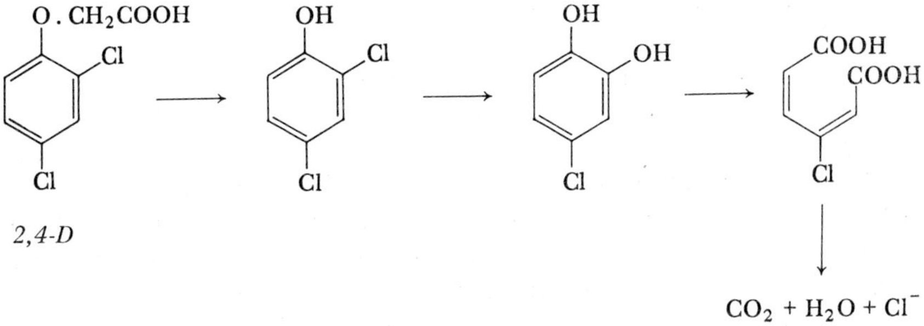

Fig. 2.16. Microbial degradation of 2,4-D by an *Achromobacter* species.

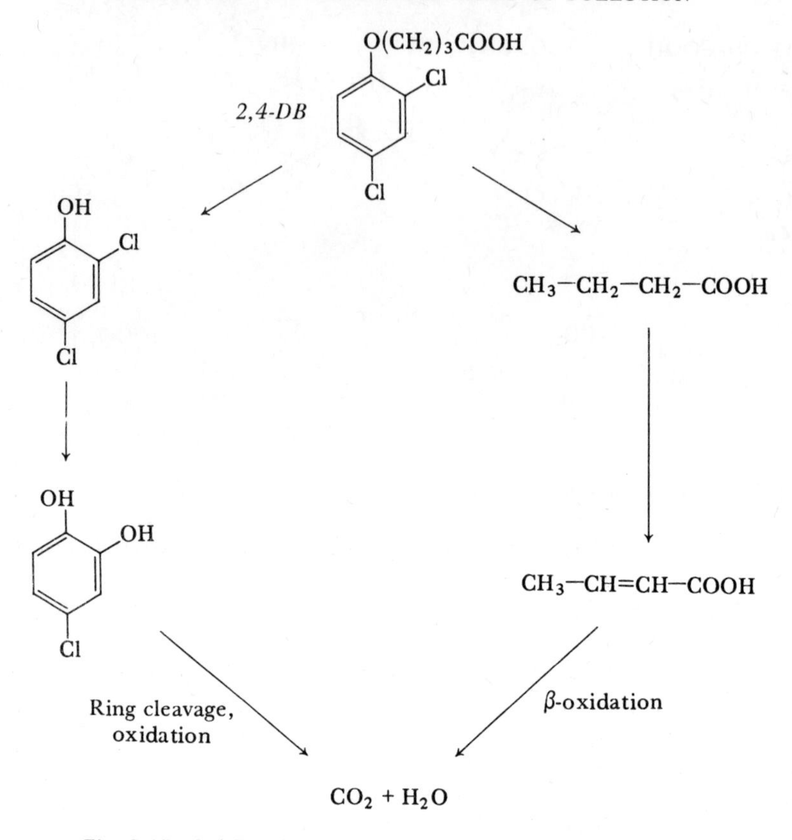

Fig. 2.17. 2,4-DB degradation by a *Flavobacterium* species.

2.6.2. Ureas. A variety of bacteria and fungi (*Pseudomonas, Xanthomonas, Sarcina, Bacillus, Penicillium* and *Aspergillus*) have been demonstrated as able to use monuron as their sole carbon source, although the details of the degradative pathway are unknown. Several other phenylureas, notably diuron, have been shown to be metabolized by microbes (Fig. 2.18).

2.6.3. Carbamates. Microbes are important in carbamate degradation in the soil; for example, IPC is totally degraded, via aniline, by *Arthrobacter* and *Achromobacter* species. It is also thought that propanil may be inactivated as shown in Fig. 2.19. The thiocarbamates (metham, diallate triallate) disappear rapidly from the soil by volatilization, but microbial decay may also be important. There is some evidence that thiocarbamates are degraded by hydrolysis followed by transthiolation and oxidation.

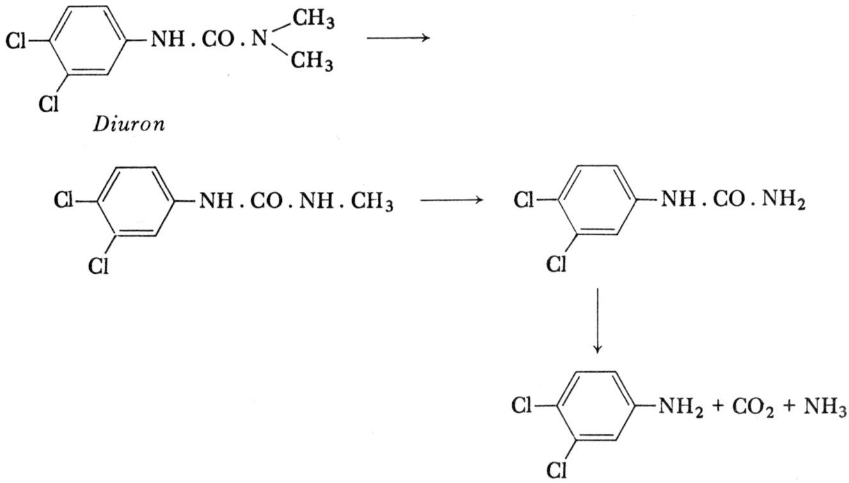

Cl, Cl — NH . CO . N(CH_3)(CH_3) ⟶

Diuron

Cl, Cl — NH . CO . NH . CH_3 ⟶ Cl, Cl — NH . CO . NH_2

↓

Cl, Cl — NH_2 + CO_2 + NH_3

Fig. 2.18. Microbial degradation of diuron.

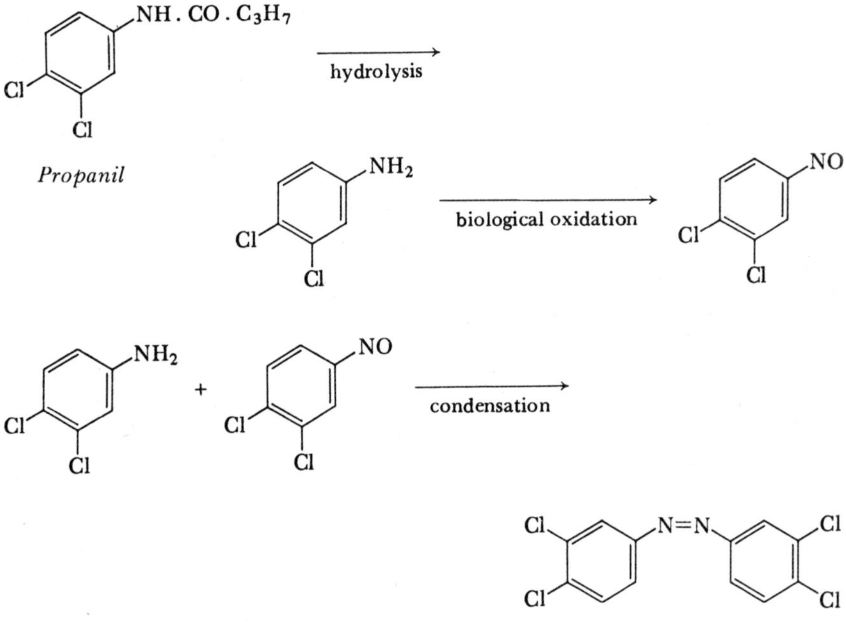

Cl, Cl — NH . CO . C_3H_7 ⟶ (hydrolysis)

Propanil

Cl, Cl — NH_2 ⟶ (biological oxidation) Cl, Cl — NO

Cl, Cl — NH_2 + Cl, Cl — NO ⟶ (condensation)

Cl, Cl — N=N — Cl, Cl

3,3,4,4-tetrachloroazobenzene

Fig. 2.19. Degradation of propanil in soil.

2.6.4. Triazines. The chlorotriazines, represented by simazine, are rapidly dechlorinated by hydrolysis in the soil (Fig. 2.20) to form their hydroxy analogues. In the main this seems to be due to

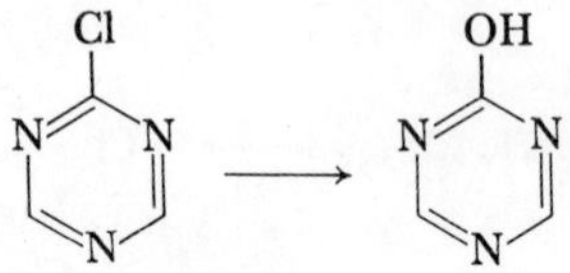

Fig. 2.20. Dechlorination of chlorotriazine.

non-biological catalysis although it has been shown that the fungus, *Fusarium roseum*, is able to hydrolyse another chlorotriazine, atrazine, to hydroxyatrazine. Triazines (Fig. 2.21), that have substituted

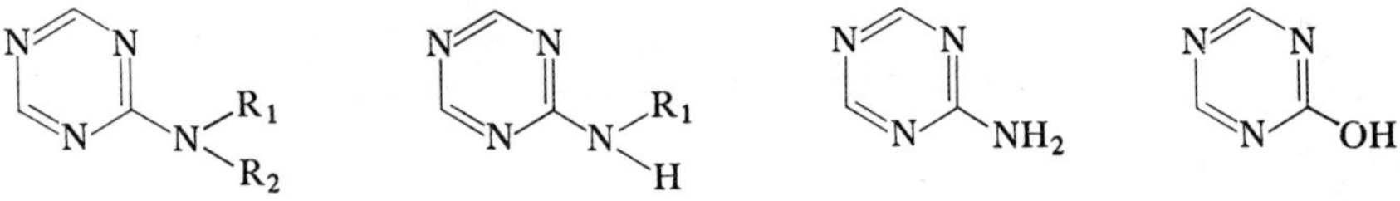

Fig. 2.21. Intermediates in triazine degradation.

amino groups in positions four or six, are microbiologically converted to their hydroxyl-derivatives(iv) after dealkylation and deammination. Dealkylation alone does not necessarily ensure detoxication. For instance, after dealkylation of the dialkylaminotriazines(i), the immediate products (ii, iii) are more toxic than the herbicide. It appears that microbes can also degrade the triazine ring but the mechanism is unknown.

2.6.5. Bipyridyls. A variety of bacteria and fungi have been shown to utilize paraquat. The early part of the degradative pathway for an unidentified bacterial isolate is shown in Fig. 2.22. The product is

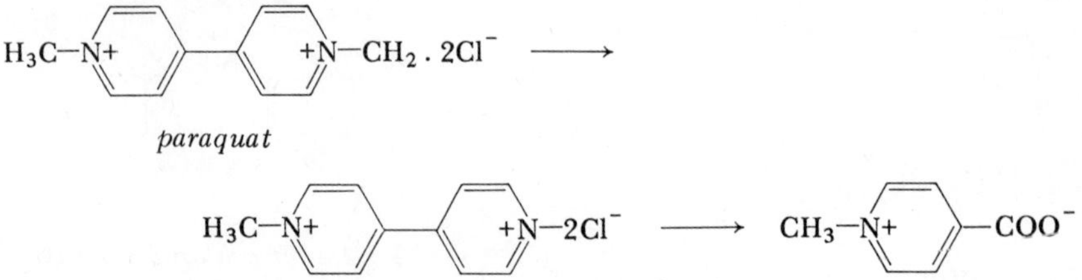

Fig. 2.22. Degradation of paraquat.

then dealkylated and the ring cleaved. A yeast species, *Lipomyces starkeyi* completely degrades paraquat and diquat and, under certain conditions, utilizes paraquat nitrogen in preference to nitrate nitrogen.

2.6.6. Toluidines. Trifluralin and related compounds are probably degraded in the soil primarily by chemical and photochemical mechanisms since degradation rates in autoclaved soils are only slightly less than in untreated soils. It is apparent, however, that in some instances microorganisms do metabolize toluidines, the initial steps involving dealkylation and reduction of nitro- and amino-groups.

2.6.7. Miscellaneous. 2,3,6-TBA is very persistent and may last for five years or more in the soil. There is some evidence for microbial degradation of this compound but if it occurs it is extremely slow. Co-metabolism has been demonstrated in the laboratory.

Amitrole is completely degraded in the soil; most of the evidence suggesting that it is due to chemical reactions and that microbes do not play an important role.

DISTRIBUTION OF PESTICIDES IN THE ENVIRONMENT

The distribution and persistence of a pesticide within the environment is a complex function of physical, chemical and biological parameters. The factors which contribute to the fate of a pesticide in soil, water and air may be considered as inherent (solubility, polarity, volatility, charge distribution, molecular size, dissociation constant) and external (adsorption, water and air movement, temperature, pH, various biological and non-biological pressures, light). Under field conditions the dissipation is often very rapid, however, if a pesticide is resistant to some or all of the forces which have the potential to attenuate its effect, it may persist for long periods of time. As a result its distribution by air and water currents may often be measured on a global rather than a local scale. In addition, the transport of pesticides in or on animal and plant material may be a subtle factor in determining the geographical range of a particular compound.

1. Soil

A superficial study of the importance of soil in determining pesticide behaviour reveals that adsorption phenomena play a key role. For example, the immobilization of a pesticide on the surface of a soil particle may (i) retard biological decay by spatially separating substrates (pesticides) and enzymes, (ii) stimulate microbial decay by concentrating enzymes and substrates, (iii) retard leaching and volatilization, (iv) catalyse non-biological breakdown, (v) fix pesticide molecules in sites of microbial proliferation and (vi) strongly influence phytotoxicity.

TABLE 2.7
Characteristics of soil inorganic components

Soil fraction	Size range in mm	Median surface area in $m^2 . g^{-1}$
Gravel	2.00–1.00	0.1
Sand	1.00–0.05	1
Silt	0.05–0.002	4.5
Clay	<0.002	115

Soil is composed of inorganic and organic fragments of various sizes and surface areas (Table 2.7). Those soil particles with the largest surface areas have the greatest potential for influencing pesticide persistence. Thus it can be deduced that of the three inorganic particles the clays are the most important. Within the clay-size fraction there are two major subdivisions, those particles with expanding lattices exemplified by montmorillonite; and those with non-expanding lattices such as kaolinite. Expanding lattice clays have both internal and external surfaces whilst those of the kaolinite type have only an external surface area (Table 2.8). Chlorite and illite clays are somewhat intermediate. As a result, within the clays themselves, montmorillonite types play a dominant role in soil/pesticide interactions.

Soil organic matter is composed of recognizable plant and animal tissues at various stages of decay and an amorphous colloidal fraction, humus. This humic material, which may contribute up to 90%

TABLE 2.8
Characteristics of soil colloids

Soil component	Surface area in $m^2 . g^{-1}$	Cation exchange capacity in meq. 100 g^{-1}
Kaolinite	25–50	2–10
Illite	75–125	15–40
Vermiculite	500–700	120–200
Montmorillonite	700–750	80–120
Colloidal organic matter	500–800	200–400

of the total carbon in many soils, is intimately associated with the inorganic fraction to form the organo-mineral complex. Like the montmorillonite clays it also has an high surface area (Table 2.8) and is of great importance in soil/pesticide interactions.

Macro-organic matter has an important effect on the distribution of persistent pesticides in food chains and is discussed elsewhere.

1.1. Adsorptive Mechanisms

The adsorption of polar or polarizable organic ions (organophosphates, carbonates, triazines, substituted areas, phenoxy-acetic acids, etc.) differs from that of inorganic ions in that it is both pH dependent and that it relies upon mechanisms in addition to simple ion exchange. A single pesticide may be adsorbed by one or more of the following mechanisms:

1.1.1. Physical. Physical adsorption is through van der Waals forces. These involve dipole-dipole interactions between pesticide moieties, on one hand, and the ionic charges in and around soil colloids, on the other. The strength of physical adsorption decreases rapidly with distance and as a result retention is greatest for those ions in close contact with the adsorbant. Thus the molecular configuration of the pesticide is important together with its susceptibility to retention by other adsorptive mechanisms. For instance, the proximity of paraquat to the clay surface is ensured by ion exchange and under these conditions, van der Waals forces also contribute to its retention.

1.1.2. Chemical. Chemical interactions result from coulombic forces which have stronger adsorptive energies than the dipole interactions involved in physical adsorption. There are at least three different types of chemical adsorption involved in pesticide/soil interactions.

1.1.2.1. Ion Exchange. The best understood mechanism of adsorption by soil particles is ion exchange. The ion exchange capacity of a soil colloid is its ability to exchange its own resident surface ions for those in solution (measured as its exchange capacity in milliequivalents 100 g $soil^{-1}$). The most important fractions involved in ion exchange, the clays and humic acids, carry a net negative charge. As a result cationic pesticides, such as paraquat and diquat, are most susceptible to adsorption by ion exchange. Soil ion exchange capacities therefore are normally measured as cation exchange capacities (CEC) (Table 2.8). Positive charges also occur, particularly at the exposed edges of clay particles, and as a result soils also have a measurable, if comparatively insignificant, anion exchange capacity (AEC).

1.1.2.2. Protonation. Pesticides which become positively charged through protonation may subsequently be adsorbed through ion exchange. Protonation involves the uptake of a proton, obtained from soil water, exchangeable H^+ on the colloid surface, or transferred from another cationic species, by a neutral molecule to give a positively charged ion. The dissociation of water, at or near a soil surface and the release of exchangeable hydrogen ions is a function of pH and therefore so is protonation. The adsorption of *s*-triazine herbicides to montmorillonite clays is reported as occurring through ion exchange mediated by protonation.

1.1.2.3. Hydrogen Bonding. Hydrogen bonding occurs when an hydrogen bridge is formed between two electronegative particles. This is a weak, low energy chemical bond involving a partial charge transfer whereas protonation involves a full charge transfer. The organophosphorus insecticide malathion is reported to be adsorbed by hydrogen bonding.

1.1.3. Co-ordination. Adsorption may involve the formation of complexes between the pesticide and the exchangeable ion of the clay particle. For instance the carbamate, EPTC, is adsorbed by ion-dipole reactions between its carbonyl group and the exchangeable metal cation of the clay. The substituted urea herbicides may also invoke this mode of adsorption.

Additional adsorptive mechanisms have been implicated in pesticide-soil relationships and include hemi-salt formation and pi bonding.

1.2. Non-Adsorptive Mechanisms

There are also some non-adsorptive interactions in soil which contribute to the persistence of certain pesticides. For example, a microorganism may assimilate pesticides which then remain intracellular and unchanged until the death of that microbe. Also the interaction of native soil organic matter and clay particles (during the formation of the organo-mineral complex) may serve to incorporate pesticides into the new structure and protect them from decay. To these must be added pesticide-lipid interactions (up to 6% of the soil humic fraction in agricultural soils is lipid), which are of particular importance in the immobilization of the non-polar chlorinated hydrocarbons.

It is apparent from this brief survey that the wide variety of immobilization mechanisms permit soil/pesticide interactions which involve neutral, anionic and cationic pesticides, those which are water soluble and those which are not and those which are polar and those which are non-polar.

2. Water

Pesticide contamination of lakes, rivers and marine environments can be attributed to the direct spraying of insecticides (to control mosquitoes, midges etc.), rain or irrigation water movement, and the discharge of waste from industries which manufacture pesticides or use them in some process such as mothproofing.

In the U.S.A. and Great Britain a range of chlorinated hydrocarbons (chlordane, heptachlor, DDT, BHC, dieldrin) have been detected in rivers. Due to their low solubility these compounds are usually associated with particulate matter and are not in solution. Thus it is not very likely that they arrived in the river through leaching but rather as a result of surface run off, spraying or association with industrial effluent. This particulate matter tends to settle out in still or slow moving waters and become concentrated in river or lake bottom mud (Table 2.9). Only in turbulent aquatic environments are pesticides transported over considerable distances.

TABLE 2.9
The distribution of DDT in a lake

Zone	No. of samples	Conc. range	Mean
Water (ppb)	82	0–22.0	0.62
Particulate matter (ppm)	33	1.8–78.0	14.74
Bottom sediment (ppm)	39	0.01–94.0	4.44

From Keith, J. O. and Hunt, E. G. (1966). Trans. 31st N. Amer. Wildlife Res. Conf. 150.

Pesticide concentrations in sea-water are rarely very high but residues may occur at river estuaries or elsewhere as a result of contaminated rainwater or dust. Even then the dilution effect is so great as to make detection virtually impossible. Nevertheless, local concentrations of pesticides attached to particulate organic matter might account for the residues found in some marine animals (fish, oysters).

The directional movement of water in terrestrial systems may be downward, upward or lateral. The soil matrix is interspersed with a complex series of pores serving as channels for solution or suspension movement. The size and shape of these pores are continuously changing making the prediction of water movement in soil, under natural conditions, very difficult. The pattern of pesticide transport in soil is further confused by the fact that both water and pesticide react with soil surfaces and that dilution with the existing soil water changes the pesticide concentration during movement. It should be noted that mass transport by water is not the only vehicle of pesticide movement and that where concentration gradients occur, ionic and molecular diffusion may be significant, at least over short distances.

A. Downward movement (leaching) is produced by capillary and gravitational forces. During the downward movement of a pesticide it will pass from the soil surface, where it is subject to volatilization, photodecomposition and co-distillation pressures, into a zone of microbial activity and adsorption. If it is resistant to attenuation by all these factors it is transported with the ground water into a stream course and possibly into the sea from where it may be carried large distances by water and air currents (as spray, particulate matter or water vapour). Movement of pesticide within the soil profile is influenced by the frequency and intensity of precipitation.

B. Upward movement, resulting from water evaporation, may cause the concentration of a chemical at the soil surface. This, in effect, returns the pesticide to zones where volatilization and photodecomposition can occur and also removes the compound from the plant root region (factors of considerable importance in determining the persistence and phytotoxicity of systemic herbicides).

C. Surface run-off from agricultural land is a source of water contamination under some conditions, especially as both soluble and non-soluble pesticides can be transported in this manner.

3. Air

With the routine use of gas-liquid chromatography it is possible to detect the varying levels of pesticide in air and rainwater. High levels of pesticide found in rainwater are usually associated with dust particles and not carried in solution (Table 2.10). It is probable that the amounts of residues in the air are only high in areas close to their application. For example, it has been estimated that aerially applied sprays are commonly less than 50% effective and much material is carried in air currents to contaminate non-target plants and animals in the vicinity.

Some have suggested that insecticides may be carried great distances in global air currents, in the same way that strontium-90

TABLE 2.10

Insecticides in air and rainwater

	Air*	Rainwater**	
G.B.	13	79.3	DDT
U.S.A.	37.3	210	DDT
G.B.	11	60.3	BHC
U.S.A.	–	23	BHC
G.B.	21	7.6	Dieldrin

From Edwards, C. A. (1970). Persistent Pesticides in the Environment, CRC Monoscience Series.

* $\mu g\ m^{-3}$

** $ng\ l^{-1}$

and cesium-137 are transported. No world wide survey, along the lines of radioactivity monitoring, has been carried out but it is likely that a high proportion of pesticide injected into the atmosphere decays as a result of temperature and sunlight before it can be deposited in rainfall. It is also unlikely that inhalation introduces significant levels of pesticide to the body.

ECOLOGICAL CONSIDERATIONS

1. Effect of Pesticides on Microorganisms

Pesticides may influence a microbial population directly by changing its metabolic and physiological activities; or indirectly by affecting plants, animals and other microorganisms. The consensus amongst research workers in the field of pesticide/microbe interactions is that, at the levels which fungicides, insecticides and herbicides reach the soil, very few long term changes occur in the microbial community. Only when pesticides are applied in large amounts and at frequent intervals are there signs of more permanent quantitative and qualitative changes.

Microbial activities assessed as indicators of pesticide response include carbon dioxide production, oxygen consumption, nitrification, growth rates and legume nodulation. If stimulation or inhibition of any of these characteristics is recorded careful interpretation is required to see if there is any relationship between *in vitro* and *in vivo* experiments, on one hand, and *in situ* conditions on the other. Some of the more obvious pitfalls are that: (i) microbial response in liquid-culture may bear little relationship to that observed in heterogenous systems such as soil; (ii) *in vivo* studies (e.g.

soil columns) do not reflect micro- and macro-climatic conditions, plant growth or animal activities; (iii) pesticide distribution in soil, aquatic systems, plants and animals is uneven and pockets of high concentration are the rule rather than the exception; (iv) the low water solubility of many pesticides invokes the use of organic solvents and it is important to gauge the individual effect of these compounds and not to confuse them with that of the solute and (v) commerical formulations usually contain carriers, wetting agents, etc. whose composition may not be revealed by the company manufacturing the pesticide and which may have some inherent biological activity which is difficult to disassociate from that of the pesticide.

1.1. Inhibition of Specific Microbes

Most work has involved the use of higher levels of pesticide (10-1000x) than those normally applied in the field. However, due to the uneven distribution of pesticides in certain environments this is not as unrealistic as it first appears. A negative response at these extreme application levels can, with some conviction, be extrapolated to mean that under similar conditions that particular pesticide will fail to produce a significant response at normal application rates. A positive response merits further examination at lower concentrations.

Many investigations have looked at the influence of pesticides on the total numbers of fungi, bacteria, actinomycetes, algae, protozoa, nematodes and arthropods in soil. Others have focussed on particular bacterial genera such as *Rhizobium, Nitrosomonas, Nitrobacter, Pseudomonas, Azotobacter* and *Thiobacillus*. Temporary restriction of fungal growth at the microenvironment level has been recorded after addition of the phenoxyacetic acid herbicides 2,4-D and 2,4,5-T to soil. High levels of insecticide may radically change the soil zoological population, destroying the highly evolved climax community and, as a result, change the comparative levels of microorganisms—especially those which had previously been on the "wrong end" of a predator-prey relationship. Fungicides, by definition, may destroy plant and animal fungal pathogens and alter the microbial flora to such an extent that it may take many months to recover. In fact, different fungal and bacterial species may dominate the new community. *Trichoderma viride*, for instance, is a significant pioneer species in the recolonization of fungicide-treated soils. The success of this microbe may, in turn, suppress the development of other fungi such as *Armillaria, Pythium, Rhizoctonia* and *Phytopthora*. On the other hand, PCNB additions may decrease competitors of *Pythium* and *Fusarium* and, as a result, increase the severity of plant disease caused by these agents.

As a group, bacterial spore formers (*Bacillus, Clostridium*) and those fungi and protozoa which form resting structures, are somewhat resistant to high levels of pesticides.

1.2. Inhibition of Microbial Transformations

It is probable that nitrification and nitrogen fixation are the most frequently assessed soil microbial activities in relation to pesticide application rates. Therefore any statements concerning the effect of pesticides on microbial transformations have to be viewed in the light of this imbalance. Nevertheless, it does appear that those micro-organisms most susceptible to pesticides are those which are involved in key biogeochemical cycles.

Nitrification is, according to some reports, inhibited by levels of propanil, ioxynil and bromoxynil less than 50 ppm. In the field, where propanil application levels are as high as 15 kg ha^{-1} (≡30 ppm), this interaction may prove important. Other herbicides (endothal, paraquat) have to be applied at levels in excess of 500 ppm for there to be any significant decline in ammonia oxidation rates. There is some evidence to sugggest that the two major chemolithotrophs involved in nitrification, *Nitrosomonas* and *Nitrobacter*, show different degrees of inhibition after triazine treatment. In general, however, herbicides are comparatively innocuous in their effect on nitrification. Fungicides, such as zineb, maneb and nabam have all been reported to decrease nitrification to some extent.

Nodulation of sweet clover may be inhibited by 50-100 ppm of aldrin and heptachlor but the degree of response is extremely varied from soil to soil. Very recently it has been suggested that organophosphorus insecticides may significantly inhibit soil enzymes such as urease. However, it is apparent from the proceeding statements that, with few exceptions, very high levels of pesticide need to be applied to soil before any major microbiological changes are induced.

2. Effect of Pesticides on Macro-Organisms

Despite the beneficial uses of pesticides there exist major dangers to Man and wildlife. Some highly toxic chemicals and their breakdown products persist in the environment for long periods of time and may move into the water of streams, rivers and seas or be carried in the atmosphere. Such is the mobility of some pesticides that no area of this planet remains free from at least some level of contamination and residues are recorded even in Antarctic zones, which are remote from areas of application. The chlorinated hydrocarbons, for instance, must now be considered an integral part of all biological systems and present in the flesh of practically all mankind.

Therefore, it is clear that over and above the intended purpose of pesticides there are side-effects which are as difficult to predict as they are potentially dangerous. Pesticides may harm non-target organisms directly, as occurs when an animal eats treated seed, or is sprayed accidently. Indirect damage presents a little less obvious relationship which an understanding of one of the basic ecological principles—the food chain—may help to explain.

Animals at the top end of chains depend on the continued success of those animals and plants lower down the hierarchy. We can use the well documented story of the chlorinated hydrocarbon insecticides to explain the importance of this sequence in terms of pesticide accumulation.

A good deal is known about the relationship of DDT with our environment, although it has taken the best part of 25 years to realize the hazards of its indiscriminate use. Chemically, DDT is a complex chlorine-containing organic molecule which is active against a wide range of insects. It kills them by disrupting their nervous system but may also stimulate or inhibit the production of enzymes and act as a carcinogen. The major physical characteristic of DDT, at least as far as animal toxicity is concerned, is its very low solubility in water and its very high solubility in lipids (fatty tissue). In consequence, because all living organisms contain lipids, DDT (and indeed other organo-chlorines such as dieldrin) tends to accumulate in animal tissues. The higher the animal is placed in a food chain the more likely it is to accumulate levels of DDT well in excess of those found in the soil, water and air of its immediate environment, a process sometimes described as biomagnification. Immobilized in this manner DDT may not cause many problems but in times of dietary stress, when reserve fatty tissue is utilized, stored insecticide is released into the blood stream.

Possibly the most dramatic example of DDT accumulation is afforded by the bird-eating Peregrine falcon. In the United Kingdom, because of the popularity of falconry, there are records of populations and nesting sites going back to the twelfth century. This population was remarkably constant for some 800 years until suddenly, in the late 1940's, it went into a rapid decline. Half a world away in California a simultaneous reduction in population size occurred and it was suspected that DDT, which was first introduced in 1945, was at least in part responsible for causing a decrease in the birth rates and survival of young birds. As mentioned previously, one of the many side effects of DDT poisoning is the stimulation of enzyme production. In birds one of the enzymes induced destroys the hormone, oestrogen. Before a bird lays her eggs her ovaries secrete oestrogen which causes her to accumulate more calcium from her diet, calcium, which is, in turn, deposited in the oviduct where it forms the eggshell. Thus a bird poisoned with DDT may produce more of this oestrogen-destroying enzyme and lay improperly calcified eggs. By measuring the thickness of falcon eggshells in museum specimens dating back to the turn of the century it was apparent that in 1947/8 a reduction of around 18% occurred. Today the Peregrine falcon is uncommon in the United Kingdom, and has disappeared almost entirely in California.

Similar decreases in eggshell thickness are observed with other birds such as the sparrow hawk which preys on other birds (especially omnivores) and lies at the end of a food chain and the national bird of the United States, the bald eagle, which lives on large fish which eat small fish which eat zooplankton which in turn eat phytoplankton. A decline in birth rate has been observed in the osprey and the Bermuda petrel, the latter illustrating the widespread nature of DDT contamination since this is a bird which rarely goes near land and the source of the DDT which it accumulates is probably the State of Arkansas more than 1000 miles away. Some brown pelicans in California are described as laying omelets (eggs with virtually no shell at all). Birds at the end of short food chains are, at present, having fewer problems whilst others, such as the seagull, appear somewhat resistant.

Another source of public concern regarding pesticide use has been the spraying of large quantities of herbicide in military defoliation programmes. In 1968 alone it was estimated that 100 000 hectares of land in South East Asia had been sprayed at a rate of approximately 32 kg ha^{-1} and that 23×10^6 kg of assorted chemicals were dumped on the countryside. Defoliation on this scale has promoted the growth of understory vegetation (bamboo) so that the original tree community may never recover. In addition, soil erosion has begun and it is possible that the rainfall pattern of large areas has been destroyed by radical changes in evaporation and transpiration rates. Moreover, the most commonly used of these herbicidal defoliants, 2,4,5-T and its impurities in commercial preparations have been shown to be teratogenic, at least to rats and mice.

Organic herbicides may also change the behavioural patterns of non-target organisms. For example, it has been reported that aphids reproduce more rapidly on broad beans treated with 2,4-D. The same herbicide makes corn more attractive to rodents and will induce cattle to eat plants which they normally avoid because of their toxicity.

CONCLUSIONS

Whilst realizing that major problems exist in the use of pesticides, few people envisage a return to the hoe and the tractor as our sole agricultural aids. The growing population demands a greater efficiency in the utilization of the decreasing areas of agricultural land. Of course the breeding of high yield and pest resistant varieties, the use of fertilizers, and the improvement in management techniques have increased production but our use of and reliance upon pesticides in agriculture is still expanding.

In any ecosystem an high proportion of potential pests are under natural biological control. The indiscriminate use of pesticides may disrupt this equilibrium and create more problems than are solved. In addition, the control of a particular pest may be only ephemeral if the causes behind its occurrence are not corrected. For example, misuse of agricultural land may lead to the establishment of a large number of weed species. The elimination of these plants (by chemical or biological control) will only be of long term value if the management of the land is improved. Similarly the destruction of disease agencies is most useful if the environment is adjusted (sanitation, drainage, etc.) so as to prevent their re-establishment.

Ultimately Man's survival will depend upon his ability to change his environment without causing adverse changes that could become self-destructive. This environment includes more than 10 000 species of injurious insects, several hundred of which are particularly destructive and need some method of continuous control; in excess of 600 species of economically important "weeds" which compete with crops for light, water, oxygen, space and minerals and some 1500 fungal and bacterial diseases. With the development of intensive agriculture, in which one crop is grown over vast areas, Man has produced situations ideal for the multiplication of these pests and predators. As a consequence he must also create efficient means of controlling them.

In the field of disease control, insecticides which eliminate the vector (carrier of disease) have been remarkably effective. In the eight years following the introduction of DDT (1945-1953) to control the mosquito it is estimated that five million lives were saved from malaria. In India alone the annual malaria mortality rate fell from 750 000 to 1500. These are the sorts of figures, in terms of the reduction of human suffering, one has to consider when criticizing the use of DDT.

In the plant world the water hyacinth chokes millions of acres of waterways throughout the world. This inhibits fish production, drives away water fowl, provides stagnant water breeding grounds for insects (especially mosquito), causes flooding and blocks drainage irrigation ditches. Economic losses without some form of control (in this case the extensive use of herbicides) would be phenomenal.

From a legal point of view food quality standards are so high that even a small percentage of disease damage renders food unsaleable. At a consumer level the public will not buy fruit and vegetables that are less than perfect. As a result the majority of food crops grown in Europe and America involve the use of pesticides during the growing period or are grown on soil which has been chemically treated in the past. Ironically it is often difficult to find untreated soils from agricultural areas for use in pesticide research.

Recommended Reading

Audus, L. J. (1964). "The Physiology and Biochemistry of Herbicides." Academic Press Inc, N.Y.

Bollag, J.-M. (1974). Microbial transformation of pesticides. *Advances in Applied Microbiology* **18**, 75.

Burns, R. G. (1975). Factors affecting pesticide loss from soil. *In* "Soil Biochemistry", vol. 4. (E. A. Paul and A. D. McLaren, Eds.). Marcel Dekker, New York.

Edwards, C. A. (1970). "Persistent Pesticides in the Environment." Butterworths, London.

Guenzi, W. D. (Ed.) (1974). "Pesticides in Soil and Water." Soil Science Society of America, Inc., Madison, Wisconsin.

Kearney, P. C., Kaufman, D. D. and Alexander, M. (1967). Biochemistry of herbicide decomposition in soils. *In* "Soil Biochemistry", (A. D. McLaren and G. H. Peterson, Eds.). Vol. 1. Marcel Dekker, New York.

Kearney, P. C. and Kaufman, D. D. (1969). "Degradation of Herbicides." Marcel Dekker, N.Y.

Matsumura, F. and Boush, G. M. (1971). Metabolism of insecticides by microorganisms. *In* "Soil Biochemistry", (A. D. McLaren and J. Skujins, Eds.), Vol. 2, Marcel Dekker, N.Y.

Mellanby, K. (1967). "Pesticides and Pollution." Fontana New Naturalist Series, Fontana, London.

White-Stevens, R. (1971). Pesticides in the Environment, Vol. 1, Part 1. Marcel Dekker, N.Y.

Woodcock, D. (1971). Metabolism of fungicides and nematocides in soils. *In* "Soil Biochemistry", (A. D. McLaren and J. Skujins, Eds.), Vol. 2, Marcel Dekker, New York.

Zweig, G. (1972). "Gas chromatographic Analysis." Academic Press Inc., N.Y.

3
Sewage and Fertilizers

The metabolic activities of Man and of his domestic animals produce vast quantities of organic waste. For the most part, regardless of whether that sewage is treated or not, its elemental components are discharged into rivers, estuaries and ultimately the sea. This may represent a long-term loss of nutrients from the environment in which we grow and nourish our food as it is depleted of essential elements—carbon, oxygen, nitrogen, phosphorus, potassium, sulphur, etc. Some, of course, have a global distribution (carbon, nitrogen, oxygen) as atmospheric components and are re-cycled to a lesser (nitrogen) or greater extent (carbon and oxygen) by plant, animal and microbial fixation. Others (potassium, sulphur, most nitrogen) need to be replaced if our agricultural land is to continue to be productive. Man has partially solved the problem by using vast quantities of organic and inorganic fertilizers which counteract this outflow.

Thus the subject of nutrient cycling in the biosphere is very much involved with the biochemistry and microbiology of both fertilizers and sewage and we consider it a rational approach to deal with these subjects in one chapter.

A. SEWAGE

INTRODUCTION

That fraction of sewage which is of domestic origin, is largely composed of metabolic wastes together with, in some countries, the macerated animal and vegetable products of kitchen disposal units. Agricultural sewage contributes faecal and urinary waste of similar composition. Industrial waste, by comparison, may contain high concentrations of metals, phenols, formaldehyde, etc., and present special problems during its subsequent treatment. In addition, sewage also contains drainage water from roads and buildings and leachate and run-off from soil.

Natural waters have the biological capacity to mineralize organic waste so long as the oxygen requirement does not exceed its availability. This process is described as self-purification during which the organic materials are converted, by aerobic microorganisms, to inorganic salts, carbon dioxide and water. The carbon dioxide is fixed by autotrophs, some of which replenish the depleted oxygen. If the biological oxygen demand (BOD) is too high (as with excessive discharge of raw sewage) anaerobic reduction of organic matter occurs with the production of foul smelling sulphides and amines. Because of this self-purification process, primitive, low population communities were able to discharge their sewage into rivers and streams without any adverse effects. Indeed the input of nutrients encouraged a large and varied aquatic flora and fauna.

With the passage of time, increasing populations, urbanization and industrialization overloaded the self-purification capacity of streams and rivers and made the disposal of waste a major problem. Some ancient cities (Assyria, Babylon) did have some form of sanitation and the Persians outlawed the discharge of organic residues into rivers. However, these are exceptions, examples of civilizations whose awareness of waste disposal problems was ahead of its time. The sanitary conditions in most European cities were appalling from the Middle Ages until well into the nineteenth century when, although the discharge of sewage into the streets was no longer allowed, many rivers could be accurately described as open sewers. Thus, in addition to the health danger (epidemics of typhoid, dysentery and cholera were common), the problems of subsequent water purification for drinking purposes were being compounded. These, and other factors, heralded the belated beginnings of sanitary engineering in the late 1880's.

TREATMENT OF SEWAGE

1. Function

From the introductory statements it is clear that the purpose of any effective sewage treatment is to remove some of the organic and inorganic matter or, failing that, to release it in an oxidized form so that it does not have a high BOD. Organic matter may be removed by filtration and sedimentation, much of the latter in the form of microbial biomass synthesized during treatment. The resulting sewage effluent will contain high levels of nitrogen and sulphur oxides and the carbon will be evolved as carbon dioxide. In addition, the effluent should be of a satisfactory microbiological standard although sewage works are not specifically designed to remove protozoal, bacterial and viral pathogens. It is perhaps fortunate that the aquatic environment in which pathogens find themselves is hostile and few survive for very long. Dilution and sedimentation also help to nullify any potential health hazards and subsequent purification for drinking purposes ensures a safe product. Most urban waste is treated although, to the concern of many, a good deal of raw sewage is still released into tidal estuaries and the sea.

2. Sewage Farms

Ideally sewage should be returned to the soil as fertilizer and not drained into the sea. The Oriental custom whereby human excrement ("night soil") was spread on the land is a theoretically sound idea but fraught with aesthetic and health problems. Composting in this manner does not kill all dangerous pathogens. At one time this principle was practised in "sewage farms" where raw effluent was run into channels and allowed to percolate into the soil. Vegetables were grown on the nutrient rich ridges between the channels. With well aerated soils, distant from water supplies, this was a reasonably safe and effective method of sewage treatment. But the soil and its microbial population have a finite capacity to deal with sewage (in the same way as the hydrosphere does) and as the area of suitable ground has decreased and volume of sewage increased this method has largely been displaced by other types of treatment.

3. Trickling Filters

Trickling filters (Fig. 3.1) and activated sludge (see 4) are the most common methods of treating sewage nowadays.

A. Raw sewage enters the sewage works and is passed through a screen to remove large pieces of insoluble material which would otherwise clog or damage pumping equipment. Some of this material may then be incinerated or converted into smaller particles for reintroduction.

B. The sewage is then passed through a grit chamber which allows sand and silt to sediment out.

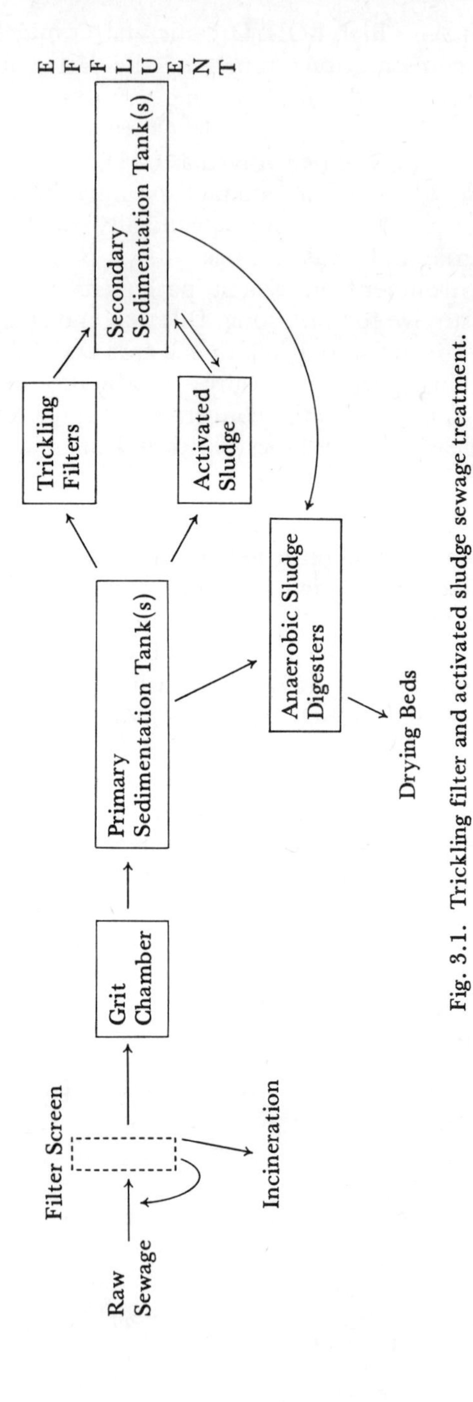

Fig. 3.1. Trickling filter and activated sludge sewage treatment.

C. Primary sedimentation tanks encourage microbial oxidation and the settling out of microbial biomass and flocculated organic matter. This material is described as raw or primary sludge. Removal of some organic matter at this stage prevents clogging of the trickling filters later on.

D. The raw sludge may then be dried and incinerated (a pollution problem in itself?) or pumped into anaerobic tanks to undergo sludge digestion. This process breaks down organic matter into organic acids, methane and carbon dioxide, the methane sometimes being used as fuel to drive the plant. The dried sludge resulting from this may be used as a fertilizer although the presence of toxic metals in this material has aroused concern in recent years (see Chapter 6). The effluent (settled sewage) is returned to the sewage flow.

E. The next step is the filtration of settled sewage through circular beds (trickling filters) containing gravel, furnace slag, corrugated plastic or any other suitable support which ensures good drainage, aeration and a surface area for microbial activity. The sewage is sprayed onto these beds from a rotating distributor and, as it flows downwards, undergoes microbial oxidation. In addition to microbes the beds are populated with larvae, metazoa and worms, all of which prey on the bacteria and help to ensure that clogging or "bulking" of the beds does not occur.

F. Any further flocculated material in the effluent is settled in "humus" tanks (the secondary sludge) and re-routed into the anaerobic digestion tanks (as in D).

G. The end-product of this treatment is an effluent containing the inorganic products of aerobic oxidation (water, salts of nitrogen, phosphorus and sulphur, and carbon dioxide) and which is suitable for release into rivers and streams.

4. Activated Sludge

In the activated sludge process the settled sewage at the end of stages C and D (above) is pumped into aeration tanks and inoculated with sludge (flocs) from an earlier batch. The mixture is stirred to ensure aeration (or pure oxygen is injected) and when the oxidation is complete, passed into sedimentation tanks. The excess activated sludge (not returned as inoculum) is anaerobically degraded in digestion tanks. The effluent from the final settlement tank is chlorinated and then discharged into the river, although advanced treatment methods may remove some of the nitrogen and phosphorus prior to release.

MICROBIOLOGY OF SEWAGE

Microorganisms isolated from sewage may arrive as one of its components (90% of the mass of faecal material is microbiological) or be involved in the various stages of its decay.

1. Microbial Components

Crude sewage contains, in addition to a large quantity of non-pathogenic microbes, a small number of agents causing infectious diseases of man. This health risk arises from the ingestion of pathogens either in drinking water or those associated with foods contaminated during growth or preparation.

Infiltration water contains mostly harmless bacteria, derived initially from soil and requiring no special treatment or monitoring. Members of this group include *Bacillus subtilis, Bacillus megaterium, Bacillus mycoides, Pseudomonas fluorescens, Achromobacter spp.* and *Micrococcus spp.* Bacteria of intestinal origin although often found in large numbers are also, for the most part, harmless and are represented by *Escherichia coli, Proteus* and *Serratia* species. Some however, such as the enterococci *(Streptococcus faecalis)* and the spore-forming anaerobes like *Clostridium perfringens* are potential pathogens, causing forms of food-poisoning. Others may promote illness of a more serious nature and their presence in sewage effluent must be continuously checked. This group is composed of three major disease-causing organisms—*Vibrio cholerae* (cholera), *Salmonella typhii* (typhoid) and *Salmonella paratyphii* (paratyphoid).

2. Microbes Involved in Sewage Decay

2.1. Anaerobic Digestion

Anaerobic digesters ferment the solids from primary sedimentation tanks and, when involved, the activated sludge process. It is essentially a two-stage reaction—one group of microorganisms hydrolysing organic matter to produce organic acids, hydrogen and carbon dioxide, whilst a second distinct group utilizes these substrates for the production of methane.

2.1.1. Non-Methanogenic. The first stage, described as non-methanogenic, is the function of a diverse group of facultative and obligate anaerobes. The species composition is determined by a variety of environmental conditions, particularly type and quantity of organic waste. The obligate anaerobes are usually the dominant bacteria with up to 1×10^9 cells ml^{-1} recorded with facultative anaerobes, occurring in densities of around 1×10^7 cells ml^{-1}. Bacteria isolated from anaerobic digesters include *Clostridium spp, Peptococcus anaerobus, Bifidobacterium spp, Desulphovibrio spp, Corynebacterium spp, Lactobacillus, Actinomyces, Staphylococcus* and *Escherichia coli*. Other physiological groups present include those producing proteolytic, lipolytic, ureolytic or cellulytic enzymes.

2.1.2. Methanogenic. The methanogenic bacteria, found in sewage digesters in quantities varying between 1×10^5 and 1×10^{10} ml^{-1} are, in many instances, similar to those isolated from the stomachs of ruminant animals and organic sediments in lakes and rivers. They are in themselves a morphologically unrelated group comprised of rods (*Methanobacterium, Methanobacillus*) and spheres (*Methanococcus, Methanosarcina*).

2.2. Trickling Filters

The trickling filter is a specialized habitat which may take many weeks to reach a steady state. Those bacteria active in the climax community and utilizing organic matter are mostly Gram negative rods such as *Achromobacter, Flavobacterium, Pseudomonas* and *Alcaligenes*. Many develop zoogloeal growth habits. The filamentous bacteria present include *Sphaerotilus natans* and *Beggiatoa*. Further down the bed profile chemolithotrophs, such as *Nitrosomonas* and *Nitrobacter*, become important members of the population.

Fungi are often in abundance and include *Fusarium*, *Penicillium*, *Mucor*, *Geotrichum*, *Sporotrichum* and a variety of yeasts. On occasions their copious growth may impede sewage flow and give rise to ventilation problems.

Algae are restricted to surface areas where light is freely available and the green colour of bacterial filters during the summer is due, amongst others, to *Phormidium, Chlorella* and *Ulothrix*.

The protozoal population varies according to depth—both in total and in species, as the substrate and levels of competition change. The commonest genera are the holozoic ciliates—particularly *Vorticella, Opercularia* and *Epistylis*. Others recorded are *Amoeba, Euglena, Persanema, Trepomonas, Paramecium* and *Stentor*.

2.3. Activated Sludge

Unlike bacterial filters, activated sludge plants have an already established microbial community as they are continuously inoculated by feed-back of activated sludge. In general terms the habitat is more homogeneous than that encountered in the bacterial filters and, as a result, there is a lower species diversity.

Bacteria are for the most part Gram negative and include members of the genera *Pseudomonas, Zoogloea, Achromobacter, Flavobacterium, Nocardia, Bdellovibrio, Mycobacterium* and the two nitrifying bacteria. Filamentous microbial growth may encourage "bulking", especially under low nitrogen conditions, the major genera responsible for this being *Sphaerotilus, Nocardia, Beggiatoa, Thiothrix, Leucothrix* and *Geotrichum.*

BIOCHEMISTRY OF SEWAGE

The major constituents of organic waste are polysaccharides, polypeptides, fats and nucleic acids and these have to be degraded to inorganic salts, carbon dioxide and water in any effective sewage treatment system. Plant, animal and indeed microbial wastes are of different composition and, as a consequence, present different substrates to the microorganisms involved in their decay. For example, animal sewage is relatively high in proteins and lipids. Plant materials, on the other hand, have an high content of cellulose and lignin. Table 3.1 give an idea of the range of substrates with which the sewage plant has to deal.

TABLE 3.1
Approximate analysis of sewage

Lipids (Ether soluble fraction)	30%
Amino acids, starch and pectins (H_2O soluble)	8%
Hemicellullose	3%
Cellulose	4%
Lignin	6%
Protein	25%
Alcohol soluble fraction	3%
Ash	21%

Most waste treatment methods in use to-day are biological processes taking advantage of the catalytic and metabolic activities of microorganisms which hydrolyse potential pollutants. The products of these reactions are ultimately converted to low molecular weight fatty acids or alcohols from which many key biochemical reactions originate. Fig. 3.2 summarizes the major reactions involved in biological waste treatment whilst detailed examples are presented in the following paragraphs.

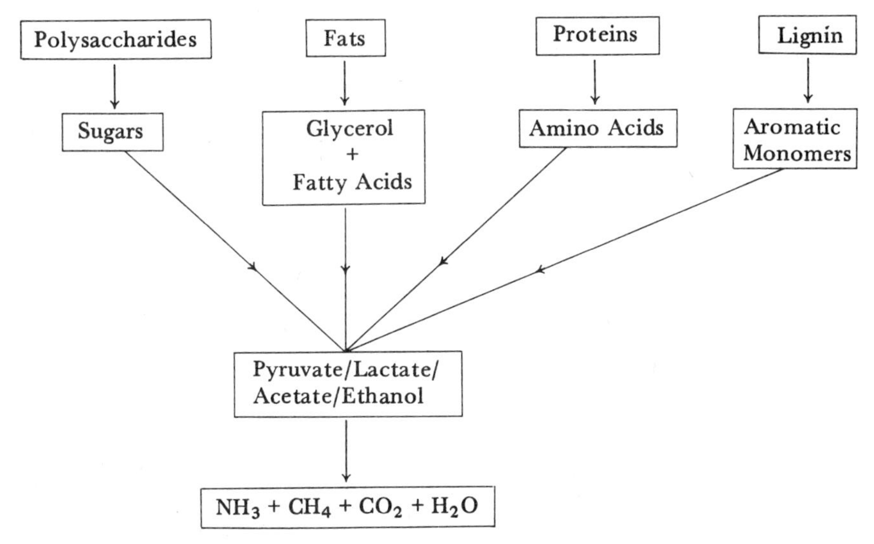

Fig. 3.2. The breakdown of organic matter during sewage treatment.

1. Polysaccharides

The carbohydrate content of living tissues is primarily in the form of polysaccharides. Their function is to provide rigidity to cell structure and to serve as reserve food sources. The bulk of dry weight in higher plants is polysaccharide whilst a smaller amount contributes to organic matter of animal origin. Both aerobic and anaerobic polysaccharide transformations occur during sewage treatment. In digestion tanks anaerobic *Clostridium* species ferment carbohydrates to produce a mixture of low molecular weight fatty acids and alcohols. The aerobic decay of the major sewage polysaccharides is described below.

Polysaccharides (or glycans) are condensation products of at least ten glycosidically-linked sugar residues. The prime function of microbes involved in their decay is to convert the large, insoluble polymer to a low molecular weight soluble fraction. This is an extracellular reaction usually involving at least two types of enzyme systems—the first splitting the glycosidic bonds to produce mostly disaccharides; the second hydrolysing the dimer to the monomer. Subsequent utilization of the monosaccharide is an intracellular event.

1.1. Cellulose

Cellulose forms the bulk of cell wall material in higher plants and, as such, is the most widespread and abundant naturally occurring organic polymer. It has been estimated that the world's reserves of cellulose are in the region of 1×10^{15} tonnes. In addition, cellulose is a minor component of animals, fungi and a very few bacteria (notably *Acetobacter xylinum*). Man has utilized cellulose in its natural form (wood) and used it in paper, cotton, rayon and many plastic products. As an indigestible component of man's diet it forms a significant proportion of human waste.

Cellulose is an insoluble, linear polymer of at least 3000 β-D-(1→4) linked glucose residues [1]. Parallel chains of polyglucans associate

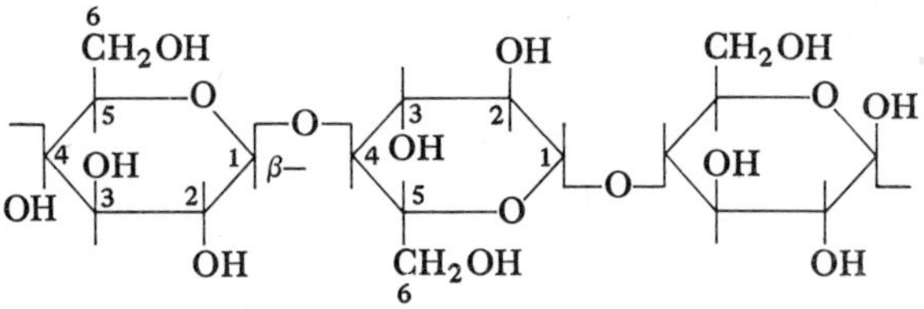

[I]

to form crystalline zones (micelles) and amorphous or paracrystalline zones. The micelles are themselves bound together to give the cellulose microfibrils typical of plant cell walls (Plate 3.1). It has been suggested that the microbial decay of this chemically and physically stable structure is the function of at least three groups of extracellular enzymes (Fig. 3.3). The first group of cellulases (C_1) may attack the cellulose fibres causing a loss of tensile strength and the production of "reactive cellulose" (sometimes called carboxymethyl cellulose or hydrocellulose). C_x cellulases perform the subsequent depolymerization to low molecular weight disaccharides (cellobiose) and oligosaccharides. The third group of enzymes catalyse the hydrolysis of these di- and oligosaccharides to glucose and are described as β-glucosidases. Where cellobiose is the substrate, cellobiase is the enzyme responsible.

With the production of soluble glucose the reaction becomes intracellular and the function of a multitude of microorganisms. Whilst most higher animals are unable to degrade cellulose (excepting those

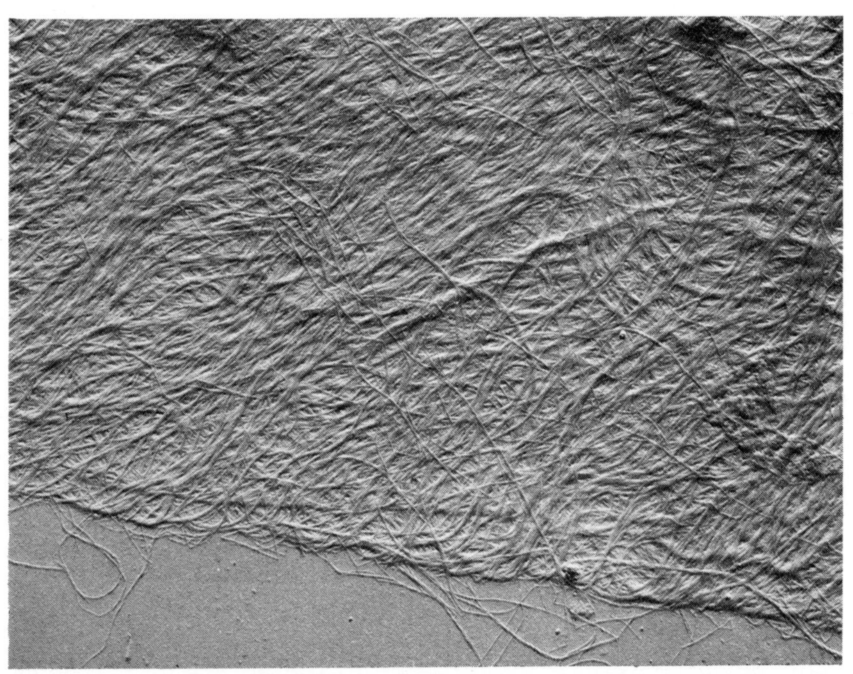

Plate 3.1. Plant cell-wall (x 15,000) showing cellulose microfibrils.
Preparation by K. Gull and R. Newsam. E.M. Unit, University of Kent.

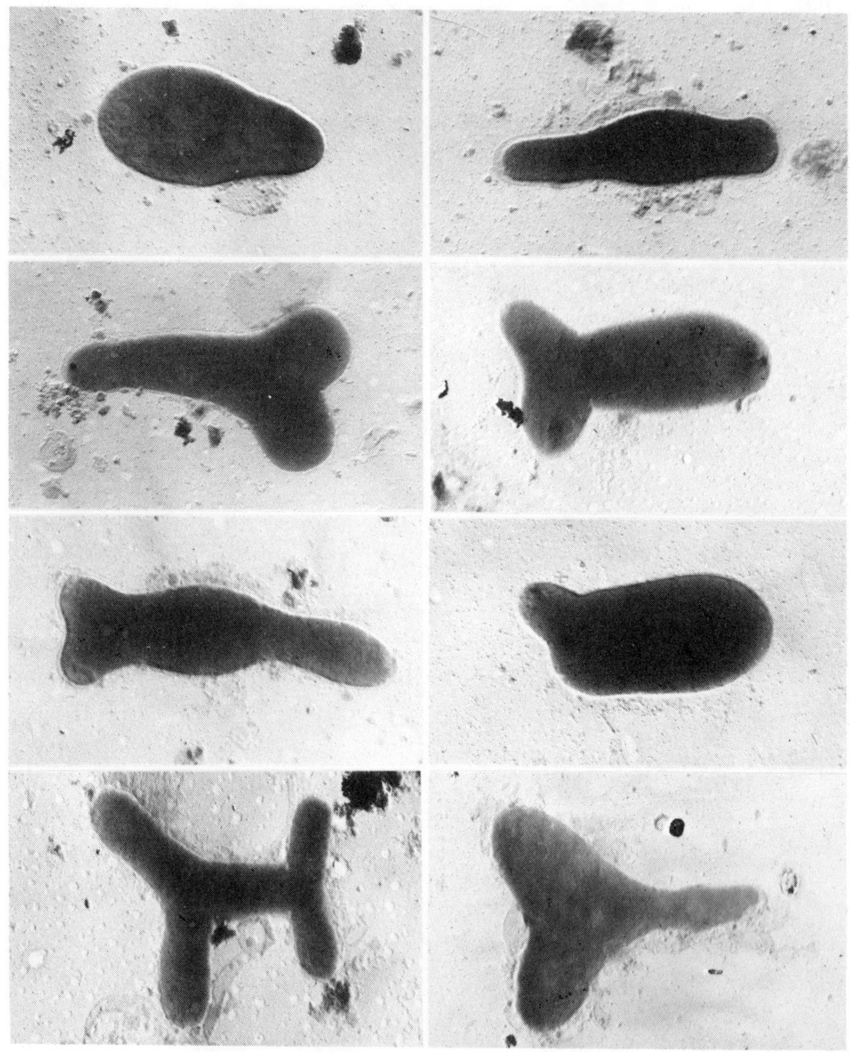

Plate 3.2. Rhizobium bacteroids (x 10,000)
Preparation by K. Gull and R. Newsam. E.M. Unit, University of Kent.

$$\text{cellulose} \xrightarrow{C_1} \text{reactive cellulose} \xrightarrow{C_x} \text{cellobiose} \xrightarrow{\text{cellobiase}} \text{glucose}$$

Fig. 3.3. Enzymic breakdown of cellulose.

having cellulytic microsymbionts) at least seventy species of bacteria, fungi, actinomycetes and protozoa have this capacity. Genera commonly reported to contain species which split cellulose include *Cytophaga, Cellulomonas, Sporocytophaga, Cellvibrio, Corynebacterium, Streptomyces, Aspergillus, Chaetomium, Cladosporium, Fusarium, Penicillium* and *Trichoderma.*

1.2. Starch

Starch is a plant food reserve polysaccharide stored either as granules or in parenchymatous tissues. It may be located in roots (tubers), stems (corms) or swollen leaf structures (bulbs) and is at its highest concentration in cereal grains. Corn or maize contains 80% starch.

Starch is composed of the linear and branched glucose polymers, amylose [II] and amylopectin, with the latter generally contributing

[II]

some 70-80% of the total. Amylose chains are constructed of some 200-300 glucose units linked in the α-D-(1→4) position. Amylopectin is an highly branched polyglucose of α-D-(1→4) linked residues and α-D-(1→6) branch points. In addition, amylopectin may contain some α-D-(1→3) linked units and some phosphate groups and is structurally related to the glycogens of the animal kingdom.

The hydrolysis of starch is catalysed by two principal enzyme systems—α- and β-amylase. α-Amylases, found in a variety of bacteria and fungi, are believed to cleave the glycosidic linkage in α-D-(1→4) glucans to produce a mixture of maltose and glucose. This, in itself, may be a two-step reaction with the intermediate production of maltotriose. It is possible that an α-glucosidase(maltase) is responsible for subsequent monosaccharide production. Certainly, the α-amylases so far investigated are far from uniform. β-Amylases perform a step-wise hydrolysis of alternate linkages and are clearly important in amylopectin decay.

A variety of other starch degrading enzymes have been described in microbiological systems. R-Enzymes, amylo-1-6 glucosidase and pullulanase all hydrolyse α-D-(1→6) bonds and cause debranching. Glucoamylase and Z-enzymes degrade starch almost completely to D-glucose whilst phosphorylases are also involved in the total pattern of decay.

1.3. Pectin

Pectins are an important group of polysaccharide materials serving as intercellular cementing substances of plant tissues and may contribute as much as 5% to the dry weight. As a consequence they are commonly encountered in sewage waste.

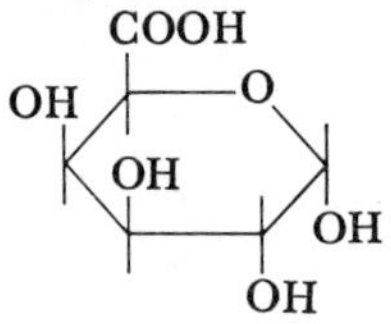

COOH
OH O
OH
OH
OH

[III]

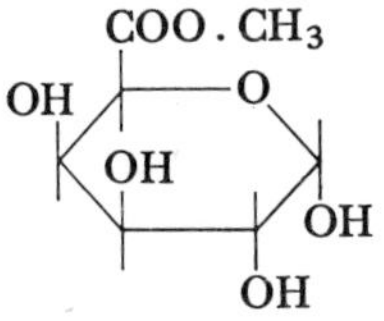

$COO.CH_3$
OH O
OH
OH
OH

[IV]

Pectins are predominantly polymers of galacturonic acid [III], methylated galacturonic acid [IV], arabinose [V] and galactose [VI]. Due to this variety of substrates two, three or even more

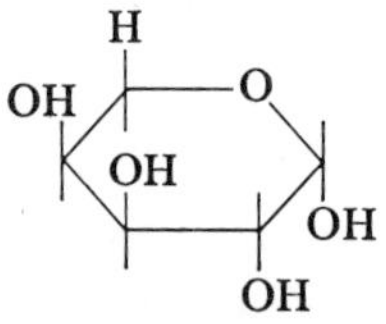

H
OH O
OH
OH
OH

(V)

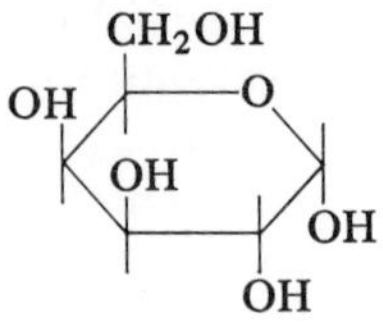

CH_2OH
OH O
OH
OH
OH

(VI)

groups of enzymes are involved in the breakdown of pectic substances. Protopectin, the parent material, is hydrolysed to form pectins or pectinic acids differing only in their physical properties. Both are polygalacturonic acids in which a high proportion of the carboxyl groups are present as methyl esters. Pectic acids, by contrast, are polyuronides which are free of methyl ester groups and may be degraded to galacturonic acids.

One group of pectic enzymes (protopectinases) solubilizes the protopectin whilst others mediate the fissure of glycosidic linkages, partially depolymerizing the substrate to form pectin. These enzymes are variously described as pectin polygalacturonase (PG), pectinase, pectolase and polygalacturonase; names which may be synonomous or representative of a multi-enzyme system. Esterases, which hydrolyze pectin and pectinic acids by splitting off methyl ester groups, are designated pectin methyl esterase (PM or PME), pectase, pectin methoxylase or pectin esterase. Finally, polygalacturonase breaks down the remaining substrate to galacturonic acid.

Pectins are readily decomposed by a variety of bacteria, actinomycetes and fungi. Amongst the bacteria are species of *Bacillus, Pseudomonas* and the "soft-rot" pathogen, *Erwinia*. Fungi include numerous molds (*Botrytis cinerea, Rhizopus, Aspergillus, Fusarium* and *Verticillium*).

1.4. Hemicellulose

Hemicellulose is a collective and somewhat imprecise term for a group of plant polysaccharide materials having no structural relationship to cellulose but from which they can be separated by alkaline extraction. Hydrolysis of the resulting material yields a complex mixture of hexose sugars, pentose sugars and frequently uronic acids. Hemicelluloses are commonly subdivided into that fraction which precipitates on acidification (hemicellulose A or cellulosans) and the supernatant (hemicellulose B or acidic hemicellulose). Hemicellulose A is predominantly composed of β-D(1→4) linked xylose units and is described as a xylan. Some molecules are linear, some highly branched and some contain L-arabinofuranose residues (arabinoxyloglycans) or traces of 4-O-methyl-D-glucuronic acid. The hemicellulose fraction of certain cereals (wheat, rye) is, for the most part, hemicellulose A. A range of additional heteropolysaccharides of the hemicellulose A type yield D-glucose and D-mannose (glucomannoglycans) or D-galactose and L-arabinose (arabinogalactoglycans) upon hydrolysis.

Hemicellulose B is an uronic acid-containing pentose polymer. Hydrolysis of this fraction reveals two major polysaccharide types. The most frequently encountered are those containing xylose and glucuronic acid whilst a second is composed of arabinose and galacturonic acid residues.

A wide variety of microorganisms are involved in the total degradation of hemicelluloses—as would be expected by the large range of substrates. Genera commonly implicated include *Bacillus, Achromobacter, Pseudomonas, Cytophaga, Lactobacillus, Vibrio, Streptomyces, Alternaria, Fusarium, Rhizopus, Chaetomium* and *Penicillium.* Within the hemicelluloses there is a good deal of specificity of microbe-substrate interaction and the microbial ecology of hemicellulose decay is clearly extremely complex. It suffices to take a look at the major component, xylan.

$$\text{Xylan} \xrightarrow{\text{xylanase}} \text{Xylobiose} \xrightarrow{\text{xylobiase}} \text{Xylose}$$

Fig. 3.4. The microbial decay of xylan.

The microbial decay of xylan (Fig. 3.4) is an extracellular two-stage process with obvious similarities to cellulose breakdown. Xylanases depolymerize xylan to the disaccharide, xylobiose, which is subsequently hydrolyzed to its monosaccharide components by xylobiase.

2. Lignin

Next to cellulose and hemicellulose, lignin is the most abundant constituent of organic waste of plant origin. Although it may contribute less than 10% of the total weight of young plants most of this is indigestible and will appear in sewage. The lignin content of mature wood varies between 20 and 40%. It has been suggested that microorganisms can repolymerize aromatic breakdown products of lignin to form humic acids. Whether this is true or not there is clearly a close chemical relationship between the components of lignin and soil humus.

Some one hundred years since the discovery of lignin its structure, synthesis and degradation are still far from fully understood. Much of this lack of knowledge arises from the difficulty of extracting a chemically unaltered lignin fraction from its physical (and maybe chemical) association with the other cell wall components (cellulose, hemicellulose, pectins).

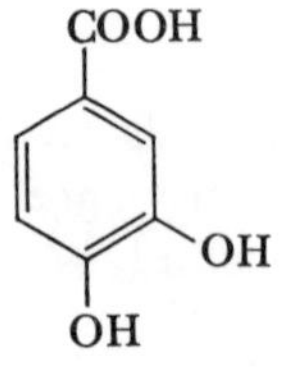

COOH
OH
OH

[VII]

Acid hydrolysis of lignin commonly yields a mixture of aromatic monomers such as protocatechuic acid [VII], *p*-hydroxybenzoic acid [VIII], vanillic acid [IX] and vanillin [X]. More extensive treatment

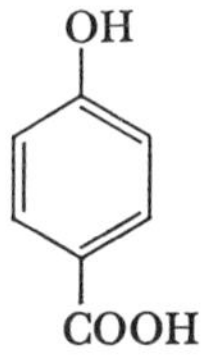

OH

COOH

[VIII]

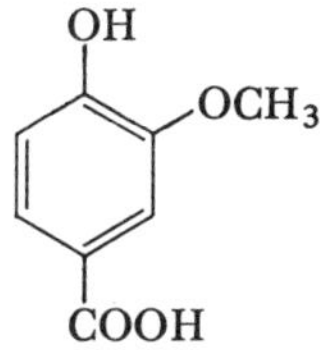

OH

OCH_3

COOH

[IX]

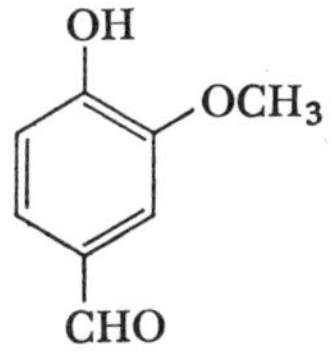

OH

OCH_3

CHO

[X]

(refluxing with ethanol and hydrochloric acid) produces what are described as "Hibbert's monomers"—α-ethoxypropionguaiacone, vanilloyl methyl ketone and others. The oxidation of lignin with nitrobenzene reveals that lignin from different sources has different components. For example, conifer lignin contains vanillin and *p*-hydroxybenzaldehyde; deciduous wood gives vanillin and syringaldehyde whilst monocotyledenous lignin contains all three. All of these aromatics are, of course, end products of harsh chemical reactions and do not necessarily indicate what complex monomers and dimers contribute to the polyphenol polymer.

It has been shown, however, that some 30% of soft-wood lignin is made up of β-aryl ester linked α-guaiacylglycerol-β-coniferyl ester units and that carbon-carbon covalent bonds may be involved in stabilizing the polymer. As a result, guaiacylglycerol has been used as a model for studying lignin decay.

The total decay of lignin is clearly the function of several groups of microorganisms—fungi, actinomycetes, bacteria. Much research, however, has involved the Basidiomycetes and especially the so-called "white-rot" fungi. Therefore, the biochemistry of lignin degradation by such as *Fomes, Armillaria, Polyporus* and *Pleurotus* is frequently

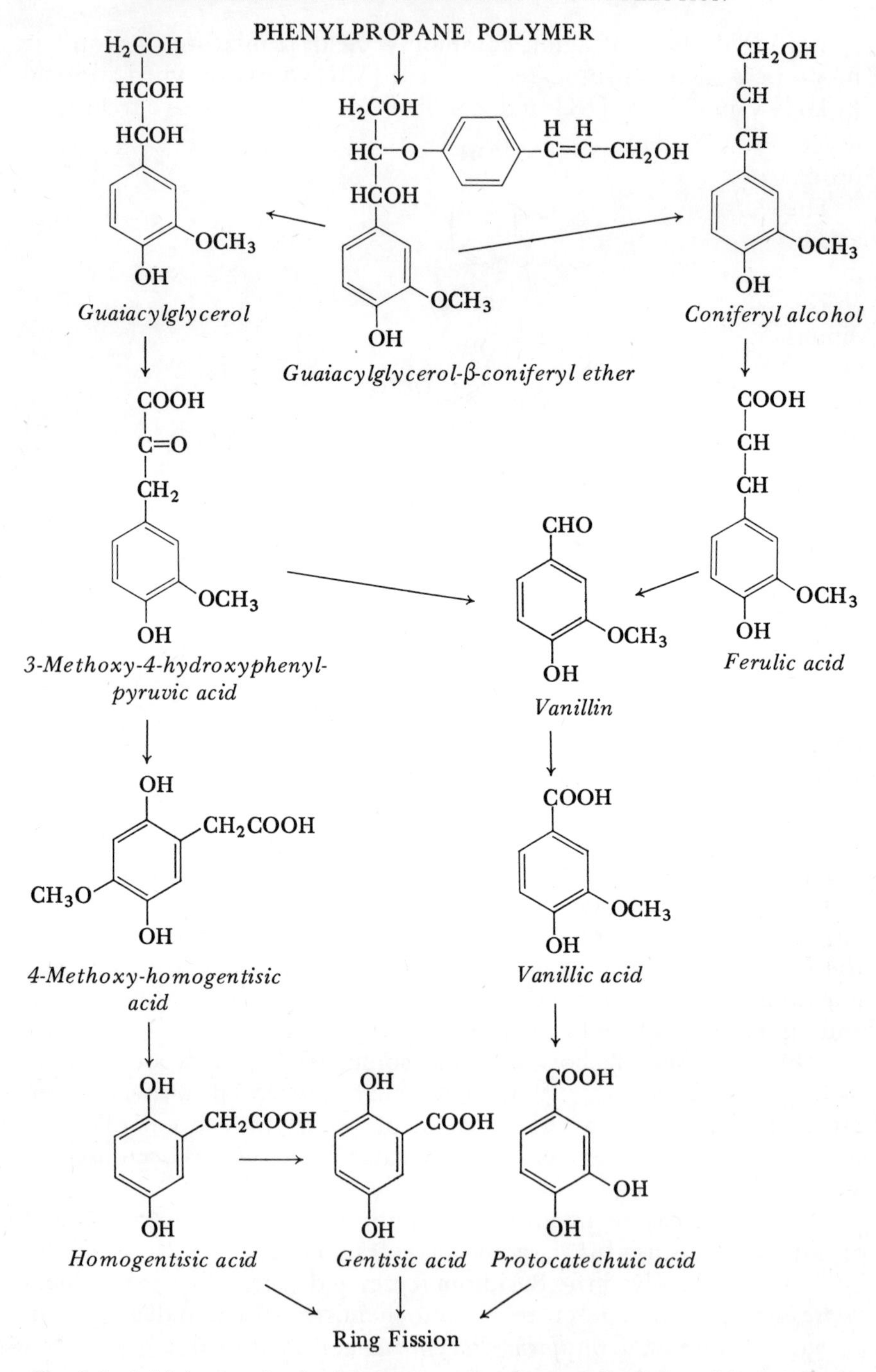

Fig. 3.5. Initial stages in the degradation of soft wood lignin by white rot fungi.

(though not satisfactorily) described. Notwithstanding, the decay of lignin in the high carbohydrate-high protein microenvironments found in sewage works must be the function of bacteria, actinomycetes and microfungi. Certainly many of the resulting aromatic products of hydrolysis are suitable bacterial and fungal substrates.

The initial steps of lignin breakdown involve the cleavage of aryl-ether linkages to depolymerise and solubilize the phenyl-propane macromolecule. De-methylation of the products (initially guaiacyl-glycerol-coniferylether) produces water soluble monomers and dimers suitable for intracellular breakdown where ring fission can occur. A number of exoenzymes are involved in the initial steps with laccase (a phenoloxidase) and peroxidase the most frequently implicated. A suggested pathway for the initial stages of lignin decay is presented in Fig. 3.5.

3. Lipids

The fat and fatty acid content of faeces is that which escapes digestion, absorption or deposition. Faecal material contains a variety of fatty substances (triglycerides, phospholipids, cholesterol, waxes and fixed oils) collectively classed as lipids. In certain pathological conditions these may constitute a substantial proportion of faeces.

The major lipid components of faeces are triglycerides (esters of higher fatty acids with glycerol) acid hydrolysis of which will give a water soluble component (glycerol) and a generally water insoluble fraction—the fatty acids.

The wide variety of fatty acids permits numerous combinations with the glyceride moiety. Fatty acids with chain lengths of 16 or 18 carbon atoms are the most common components of fats of animal and vegetable origin—the saturated acids, palmitic ($C_{15}H_{31}COOH$) and stearic ($C_{17}H_{35}COOH$) and the unsaturated acid oleic ($C_{17}H_{33}COOH$). These and other commonly occurring fatty acids (Table 3.2) originate from plant and animal foods and, as such, form a part of man's diet (and, as a consequence, his faeces).

In addition, fats and fatty acids of bacterial and fungal origin are not uncommon especially in environments where carbohydrate levels are high. Examples include acetic (CH_3COOH), propionic (C_2H_5COOH), butyric (C_3H_7COOH), valeric (C_4H_9COOH), lauric ($C_{11}H_{23}COOH$) and a range of branched chain fatty acids (6-methyloctanoic, tuberculostearic, phytomonic and phthioic).

Lipase enzymes, common in plants and animals, are found in both aerobic and anaerobic microorganisms. Enzymic hydrolysis of fats yields a mixture of glycerol and fatty acids (Fig. 3.6).

$$
\begin{array}{l} CH_2O.CO.R_1) \\ | \\ CH.O.CO.R_2) \\ | \\ CH_2O.CO.R_3) \end{array}
\xrightarrow{\text{lipase}}
\begin{array}{l} CH_2OH + R_1{-}COOH \\ | \\ CHOH + R_2{-}COOH \\ | \\ CH_2OH + R_3{-}COOH \end{array}
$$

Fig. 3.6. Fat hydrolysis.

Long-chain fatty acids are subsequently metabolized by β-oxidation in which two carbon units at a time are split from the chain.

TABLE 3.2
Fatty acids in foods

(a) *Saturated Fatty Acids*			(b) *Unsaturated Fatty Acids*		
Butyric	C_3H_7COOH	milk	Myristoleic	$C_{13}H_{25}COOH$	seafish
Caproic	$C_5H_{11}COOH$	milk, coconut	Palmitoleic	$C_{15}H_{29}COOH$	seafish
Caprylic	$C_7H_{15}COOH$	milk, coconut	Oleic	$C_{17}H_{33}COOH$	numerous
Capric	$C_9H_{19}COOH$	milk, coconut	Gadoleic	$C_{19}H_{37}COOH$	seafish
Lauric	$C_{11}H_{23}COOH$	coconut	(c) *Others with one or more double bonds in chain*		
Myristic	$C_{13}H_{27}COOH$	nutmeg			
Palmitic	$C_{15}H_{31}COOH$	numerous	Petroselinic	$C_{13}H_{25}COOH$	milk
Stearic	$C_{17}H_{35}COOH$	numerous	Vaccenic	$C_{17}H_{33}COOH$	milk
Arachidic	$C_{19}H_{39}COOH$	peanuts, milk	Linoleic	$C_{17}H_{31}COOH$	many plants
Lignoceric	$C_{23}H_{47}COOH$	peanuts	Linolenic	$C_{17}H_{29}COOH$	many plants

4. Proteins

Proteins are long-chain polymers of L-amino acids with the general structure $H_2NCHRCOOH$ where R may be a hydrogen atom, a single methyl group, a short carbon chain (with or without functional groups such as SH, NH_2 and COOH) or a cyclic structure. The amino acids are joined by peptide bonds (CO–NH).

Exoenzymes which hydrolyze the peptide bonds are collectively described as proteases and appear to be of two major types; exopeptidases which cleave peptide bonds towards the end of the protein chain, and endopeptidases which function distant from terminal amino acids. Proteolytic activity produces a mixture of peptones (long chains of amino acids), peptides (short chains) and free amino acids. There is evidence that some proteases are specific for linkages between certain amino acids. For example, subtilisin (from *Bacillus subtilis*) will only react with serine linkages. Common proteolytic genera are *Pseudomonas*, *Bacillus* and *Micrococcus*.

Amino acids serve as carbon and nitrogen sources for numerous heterotrophs and their intracellular deamination and decarboxylation is well documented. Some of them are metabolized readily (arginine, tryptophan) whilst others are more resistant (lysine, methionine). Proteins and amino acids may also be attacked fermentatively (*Clostridium* spp) to produce a number of unpleasant smelling amines, mercaptans and hydrogen sulphides. This process is known as putrefaction.

The microbial decay of other nitrogen-containing organic materials is outlined elsewhere in this chapter.

5. Methanogenic Bacteria

One of the major groups of microorganisms engaged in the anaerobic digestion of sludge is the methanogenic bacteria. These organisms couple organic oxidation with carbon dioxide reduction to produce methane. The substrates used by methanogenic bacteria fall into three distinct categories.

A. The lower fatty acids containing six or less carbon atoms (formic, acetic, propionic, butyric, valeric, caproic).

B. The normal and iso-alcohols containing from one to five carbon atoms (methanol, ethanol, propanol, butanol, pentanol).

C. Three inorganic gases (hydrogen, carbon monoxide and carbon dioxide).

In addition microbial species may be highly specific for one or more of these substrates (Table 3.3). Methane bacteria are intolerant

TABLE 3.3
Methanobacterium species and their carbon substrates

Species	Substrate
M. formicicum	CO, CO_2, H_2
*M. omelianskii**	alcohols
M. suboxydans	butyrate, valerate, caproate
M. vanielli	formate, H_2
M. barkerii	methanol, acetate, formate

*A two-membered association.

of pH extremes (optimum pH 6.4-7.2) and this parameter must be monitored in sewage digesters. For example, if the input of carbohydrates is high compared to proteins there is relatively little available ammonia (nitrogen source for microbes) and more carbon dioxide than nitrogen is produced. This will cause a pH drop. By contrast, high protein input results in too much ammonia production and resulting alkalinity.

The present state of knowledge indicates that methane can be formed in two different ways (Fig. 3.7):

A. With some substrates (ethanol (i), butyrate, hydrogen (ii)) methane is produced by atmospheric carbon dioxide reduction. Removal of carbon dioxide stops the reaction.

B. Methane is formed by carbon dioxide reduction when that gas is formed during substrate oxidation (carbon monoxide (iii), propionate (iv), acetate (v)).

(i) $2C_2H_5OH + CO_2 \longrightarrow 2CH_3COOH + CH_4$

(ii) $4H_2 + CO_2 \longrightarrow CH_4 + 2H_2O$

(iii) $CO + H_2O \longrightarrow CO_2 + H_2$

$CO_2 + 4H_2 \longrightarrow CH_4 + 2H_2O$

$CO + 3H_2 \longrightarrow CH_4 + H_2O$

(iv) $4C_2H_5COOH + 8H_2O \longrightarrow 4CH_3COOH + 4CO_2 + 24H$

$3CO_2 + 24H \longrightarrow 3CH_4 + 6H_2O$

$4C_2H_5COOH + 2H_2O \longrightarrow 4CH_3COOH + CO_2 + 3CH_4$

(v) $CH_3COOH \longrightarrow CH_4 + CO_2$

Fig. 3.7. Formation of methane by methanogenic bacteria.

Both mechanisms appear to involve the reduction of carbon dioxide to N^5, N^{10}-methylenetetrahydrofolate and thence, via methyl cobalamine and coenzyme-M, to methane. The electrons involved in this reaction are derived from either the organic substrate or the hydrogen.

ANALYTICAL METHODS

The efficiency of sewage treatment is routinely measured by physical, biological and chemical examination of the effluent. Care needs to be taken in sampling as effluent varies almost hourly

1. Physical Measurements

A. Colour: Satisfactory effluents are generally a pale straw colour. Any variation from this is not necessarily a result of inadequate treatment but may be caused by industrial waste, algal growth and other factors.

B. Odour: Sewage effluent of domestic origin should be essentially odourless and certainly not smelling of ammonia or hydrogen sulphide. Industrial effluent may have a characteristic oily, soapy or aromatic odour.

C. Reaction: Most sewage effluents are neutral to alkaline although high ammonia levels will increase that alkalinity whilst industrial components often give rise to an acid effluent.

D. Suspended Solids: After filtering and drying of a sample the solid effluent should not exceed 30 $mg . l^{-1}$.

2. Chemical Measurements

A. Inorganic nitrogen: This employs a variety of methods to measure ammonia, nitrite and nitrate levels. Generally speaking, it is more acceptable for the nitrogen to be in its most oxidized form.

B. Organic nitrogen: Estimated by Kjeldahl digestion.

3. Biological Measurements

The three commonly used biological measurements all test the oxygen requirements of the sewage effluent (for the reasons described previously—see page 64).

A. Permanganate value (PV): This measures the amount of oxygen absorbed by permanganate in four hours at 27°C. "Clean" sewage effluent should register less than 10 $mg . l^{-1}$.

B. Chemical oxygen demand (COD): This method records the maximum potential adsorption capacity of the organic matter in a sample. The sample is boiled with acidic potassium dichromate for two hours after which any unreacted dichromate is determined by titration with ferrous ammonium sulphate. In some instances COD levels may be five or more times that recorded for BOD implying that sewage effluent contains a considerable amount of organic matter not readily attacked by microbes.

C. Biochemical Oxygen Demand (BOD): BOD is a measure of the amount of dissolved oxygen absorbed by a test sample held at 20°C over a period of five days. The dilution factor must be such that all the oxygen is not consumed in the time period. Suitable dilutions range from 1 in 5 for "clean" effluents to 1 in 50 or more for inadequately treated effluent. COD and PV measurements help in determining the dilution factor. One method of classifying rivers based on BOD measurements is shown in Table 3.4.

TABLE 3.4
Classification of rivers according to BOD

Classification	O_2 absorption (in mg . l^{-1} per 5 days)
Very clean	1
Clean	2
Fairly clean	3
Doubtful	5
Polluted	10

BOD measurements are not totally satisfactory for a number of reasons, such as the presence of microbial inhibitors in the effluent, sampling errors or whether oxygen consumption by nitrifiers should be included.

Analyses of sewage outflow cannot be considered complete without further measurements of phosphate, chlorine, pesticides (Chapter 2), cyanide, toxic metals (Chapter 7), detergents (Chapter 5), microbial pathogens and many other parameters.

B. FERTILIZERS

INTRODUCTION

Modern agriculture relies on four major technologies to produce the large quantities of food demanded by an expanding population. These are mechanization, irrigation, fertilization and pest control. In this section we shall examine the development and use of chemical fertilizers and some of the ecological problems arising from this use.

In the nineteenth century the German chemist Justus von Leibig analysed the mineral or ash content of plants and published information on the comparative occurrence of various elements in plant tissues. This led him and others to suggest that if a soil contained all these elements, in the correct proportions, optimum plant growth would be supported. Indirectly, the first inorganic fertilizer had been described.

The burning of organic material was then, and still is, the standard approach to discovering its chemical composition. The products of combustion are mostly carbon, oxygen and hydrogen (as CO_2 and H_2O) with much smaller proportions of nitrogen (0.5% by wt). A "typical" plant analysis appears in Table 3.5.

TABLE 3.5
Ash analysis of a *Zea Mays* plant*

Element	% of total dry weight	% of total ash
Nitrogen	1.46	25.9
Silicon	1.17	20.8
Potassium	0.92	16.3
Calcium	0.23	4.1
Phosphorus	0.20	3.6
Magnesium	0.18	3.2
Sulphur	0.17	3.0
Chlorine	0.14	2.5
Aluminium	0.11	2.0
Iron	0.08	1.4
Manganese	0.04	0.7
Others	0.93	16.5

* From Sutcliffe, J. F. "Mineral Salt Absorption in Plants" 1962, Pergamon.

The first artificial fertilizers used in the mid-nineteenth century were mixtures of guano and potash, but it wasn't until the twentieth century that the widespread use of inorganic fertilizers became necessary due to shrinking agricultural areas, declining soil fertility, and expanding populations. The first countries to feel these pressures (and thus adopt the new agriculture) were Japan, the Netherlands, Denmark, Sweden and the U.S.A. Experience has now shown that the application of chemical fertilizers in conjunction with other management changes, may double, triple or even quadruple yields. Nowadays chemical fertilizer production is a multi-million dollar industry.

It is apparent that the two most common elements limiting plant growth are nitrogen and phosphorus.

NITROGEN

Nitrogen, as a key component of nucleic acids and proteins, is an element essential to all living things. It is, however, in many instances, available only at sub-optimal levels. By availability we mean the nitrogen which at any one time is in a form that is readily absorbed by the plant roots (usually as nitrate). Clearly, the overall levels of nitrogen in the biosphere are well in excess of plant and animal requirements.

It is useful at this stage to outline the major microbiological and biochemical mechanisms by which an unavailable form of nitrogen may be transformed to an available one during the course of the nitrogen cycle.

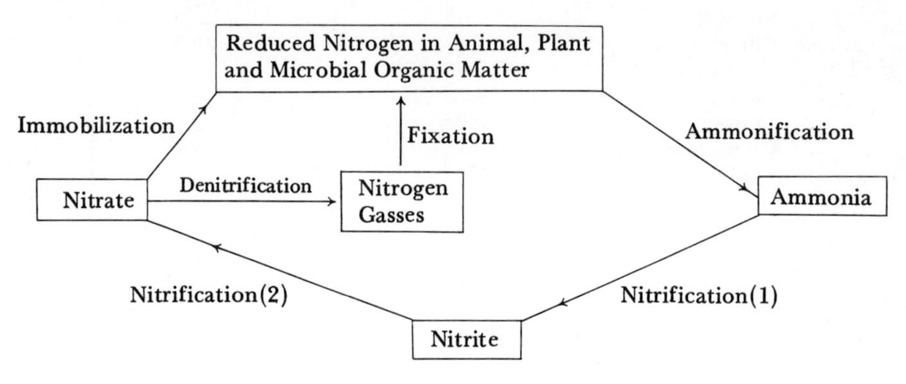

Fig. 3.8. The nitrogen cycle.

The supply of nitrogen to higher plants is almost entirely controlled by terrestrial and aquatic microorganisms. There are five distinct but inter-related components of the nitrogen cycle (Fig. 3.8)—ammonification or mineralization, nitrification, immobilization, fixation and denitrification.

1. Ammonification

Plant and animal debris is degraded by a range of microorganisms which utilize the carbon, nitrogen and energy produced for cellular growth and division. Any carbon over and above the microbial population's direct requirement is ultimately released as carbon dioxide whilst any nitrogen appears as ammonium compounds or is evolved as ammonia.

A large number of aerobic and anaerobic heterotrophs are capable of performing this reaction and, at any one time, one gram of soil may contain as many as 1×10^8 bacteria, fungi and actinomycetes all with the potential to ammonify a suitable substrate such as proteins, nucleic acids, amino-acids, sugar-amines, calcium cyanamide or urea.

1.1. Nucleic Acids

Nucleic acids reach the soil from decomposing plant, animal and microbial biomass. There are two types–ribonucleic acid (RNA) and deoxyribonucleic acid (DNA) each being composed of a chain of nucleotides containing a pentose sugar unit, a purine or pyrimidine base (Fig. 3.9) and a phosphoric acid residue. Thus in one organic substrate we have a starting point for carbon, nitrogen and phosphorus cycles.

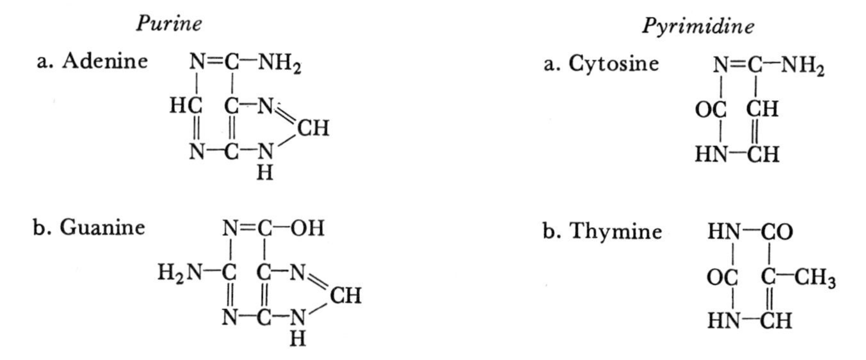

Fig. 3.9. Nucleic acid bases.

The degradation of nucleic acids is a sequential extracellular reaction. The large polynucleotides are converted to smaller fragments and eventually mononucleotides. The two enzymes responsible for this are ribonuclease and deoxyribonuclease which attack RNA and DNA respectively and are synthesized by a large number of microbial genera such as *Penicillium, Aspergillus, Streptomyces, Clostridium* and *Bacillus.* The next steps involve removal of sugar and phosphate units (involved in C and P cycles) and the intracellular decay of the purine and pyrimidine bases.

1.2. Urea

Urea is a product of the destruction of nucleic acid bases, an important agricultural fertilizer and an excretory product of animals and, as such, a key component of the nitrogen cycle. Many microorganisms possess the enzyme urease and, in addition, considerable extracellular ureolytic activity is associated with soil humic material. Ammonium carbonate is an intermediate in urea hydrolysis (Fig. 3.10).

$$CO(NH_2)_2 + H_2O \longrightarrow NH_4CO_2NH_2 \longrightarrow 2NH_3 + CO_2$$

Fig. 3.10. Urea hydrolysis.

Most frequently studied urea degrading bacteria include species of *Bacillus, Micrococcus, Pseudomonas, Achromobacter, Corynebacterium* and *Clostridium*, together with filamentous fungi and actinomycetes. The ability to hydrolyse urea is a diagnostic test for "Group 3" *Bacillus* species.

1.3. Calcium Cyanamide

Calcium cyanamide is an important nitrogenous fertilizer which is converted rapidly in soil to ammonia. The biochemistry involves three distinct processes (Fig. 3.11)—the hydrolysis of calcium cyanamide to cyanamide (i), conversion of cyanamide to urea (ii) and subsequent microbial hydrolysis (iii). Steps (i) and (ii) may be non-biological.

$$\text{(i)}\quad CaCN_2 + 2H_2O \longrightarrow H_2CN_2 + Ca(OH)_2$$

$$\text{(ii)}\quad H_2CN_2 + H_2O \longrightarrow CO(NH_2)_2$$

$$\text{(iii)}\quad CO(NH_2)_2 + H_2O \longrightarrow 2NH_3 + CO_2$$

Fig. 3.11. Conversion of calcium cyanamide to ammonia in soil.

1.4. Ecological Factors

Some nitrogen-containing organic substrates are more easily degraded than others. For example, those residues containing an abundance of non-nitrogenous organic matter (cellulose) may encourage the development of a high proportion of microbes capable of utilizing glucose polymers but not able to ammonify any nitrogen-containing organic matter. The presence of naturally occurring bacteriostatic phenolics and ligno-protein complexes may also retard ammonification.

Ammonification is both an aerobic and an anaerobic process and thus the displacement of oxygen by water does not inhibit ammonia production to any great extent. Simply, other microorganisms become involved and the end products (in addition to CO_2 and NH_3) are different (Fig. 3.12). However, as the next step in the nitrogen cycle—nitrification—is obligately aerobic, anaerobic environments do inhibit the overall cycle.

Environment pH also influences the rate of ammonification which appears most efficient in alkaline/neutral environments. However,

$$N\text{-organic matter} \xrightarrow{\text{aerobic}} NH_4^+ + CO_2 + SO_4^{2-} + H_2O$$

$$N\text{-organic matter} \xrightarrow{\text{anaerobic}} NH_4^+ + CO_2 + \text{amines and organic acids}$$

Fig. 3.12. Ammonification.

even in acid environments fungal proteases ensure some turnover of protein material.

2. Nitrification

2.1. Microbiology and Biochemistry

Nitrogen, in its reduced form ammonia, is the starting point for a nitrification process which has two distinct steps (Fig. 3.13): oxidation of ammonia to nitrite (i) and conversion of nitrite to nitrate (ii). Two obligately aerobic chemolithotrophic bacterial genera are prominent in this process—*Nitrosomonas,* which performs the first reaction, and *Nitrobacter* which performs the second.

$$\text{(i)}\quad 2NH_4^+ + 3O_2 \rightarrow 2NO_2^- + 2H_2O + 4H^+ \qquad \Delta F(\text{kcal}) - 84{\cdot}0$$

$$\text{(ii)}\quad 2NO_2^- + O_2 \rightarrow 2NO_3^- \qquad \Delta F(\text{kcal}) - 17{\cdot}8$$

$$2NH_4^+ + 4O_2 \rightarrow 2NO_3^- + 2H_2O + 4H^+ \qquad \Delta F(\text{kcal}) - 101{\cdot}8$$

$$\text{(iii)}\quad 2NH_4^+ + O_2 \rightarrow 2NH_2OH + 2H^+$$

$$\text{(iv)}\quad NH_2OH + HNO_2 \rightarrow NO_2NHOH + 2H$$

$$\text{(v)}\quad 2NO_2NHOH + O_2 \rightarrow 4HNO_2$$

$$\text{(vi)}\quad HNO_2 + H_2O \rightarrow HNO_3 + 2H$$

Fig. 3.13. Nitrification.

The detailed biochemistry (Fig. 3.13) of these reactions is poorly understood. It appears that the first stage in nitrification is the oxidation of ammonia to hydroxylamine (iii). Following this the hydroxylamine may react with nitrite already formed (as nitrous acid) to produce nitrohydroxylamine (iv) which is subsequently oxidized to nitrite (v). *Nitrobacter* oxidizes nitrite to nitrate directly (vi).

2.2. Ecological Factors

There are a large number of physical and chemical factors which influence the nitrification process, some of which will be indirect in that they affect mineralization rates.

If organic materials in the soil are low in nitrogen (and microbial demand outstrips supply) then plant-available nitrate will be limited. In addition, if microbial oxygen consumption is high then the resulting anaerobic conditions will prevent ammonia oxidations. The same retardation occurs under high water levels when oxygen is excluded from the soil pores.

It is apparent that during the nitrification process hydrogen ion production tends to acidify microbial environments which the soil buffering capacity normally counteracts. This is just as well because the nitrifying organisms, unlike the ammonifiers, are very sensitive to acidity. *Nitrobacter* species have an optimum pH of between 5 and 8, whilst *Nitrosomonas* functions best between 7 and 9. Therefore in slightly acid environments *Nitrobacter* soon runs out of nitrite substrate whilst at alkaline reactions there may be a build up of the biologically toxic nitrite. At a pH of less than 5 nitrification stops all together.

Soil colloidal clays may strongly adsorb ammonium ions and temporarily restrict their input into the nitrification pathway.

3. Denitrification

3.1. Microbiology and Biochemistry

As aerobic conditions stimulate the microbial oxidation of ammonia so, under anaerobic conditions, reduction of inorganic nitrogen occurs.

Denitrification has been recognized as a source of soil nitrogen losses for decades and refers to the biological reduction of nitrate to nitrite, nitrous oxide, nitric oxide, ammonia and gaseous nitrogen. Some of these products are re-incorporated into the soil environment (NO_2, NH_3) and as a result are only temporarily removed from the immediate nitrogen cycle. Others are released in a gaseous form (N_2 NO, N_2O) into the super-soil atmosphere, and should be considered as a more permanent loss. The former is often referred to as nitrate reduction, the latter as true denitrification.

A few autotrophic bacteria are capable of reducing nitrate by using it as an electron acceptor. Examples of these include *Micrococcus denitrificans* and the sulphur bacterium *Thiobacillus denitrificans* (Fig. 3.14).

$$5S + 6KNO_3 + 2CaCO_3 \rightarrow 3K_2SO_4 + 2CaSO_4 + 2CO_2 + 3N_2$$

Fig. 3.14. Nitrate reduction by *Thiobacillus denitrificans.*

However, the great majority of nitrate-reducing bacteria are facultatively anaerobic heterotrophs and include species of the genera *Pseudomonas, Achromobacter, Micrococcus* and *Bacillus.* The reducing power comes from glucose (Fig. 3.15) and for each two molecules of nitrate one molecule of nitrous oxide or nitrogen is produced.

$$C_6H_{12}O_6 + 6KNO_3 \rightarrow 6CO_2 + 3H_2O + 6KOH + 3N_2O$$

or

$$5C_6H_{12}O_6 + 24KNO_3 \rightarrow 30CO_2 + 18H_2O + 24KOH + 12N_2$$

Fig. 3.15. Nitrate reduction by heterotrophic bacteria.

Heterotrophic denitrifiers are often in great abundance in soil ($1 \times 10^{4-6}\ g^{-1}$) performing aerobic activities until oxygen becomes scarce. Thus the *potential* for denitrification is enormous.

The most likely pathways for nitrate reduction and denitrification appears in Fig. 3.16.

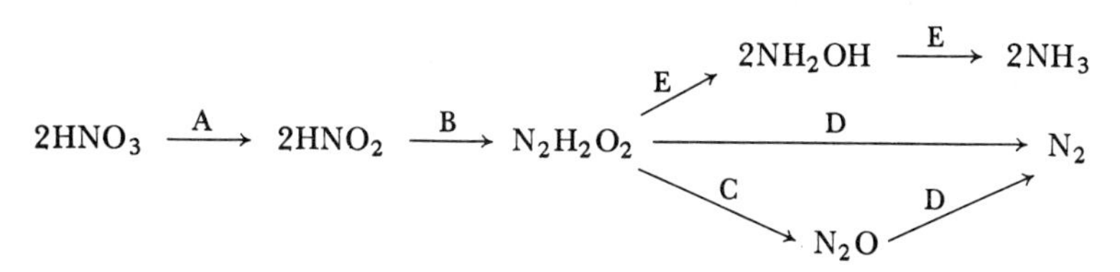

Fig. 3.16. Pathways of nitrate reduction and denitrification.

This reaction is composed of the following steps:

A. A reduction involving the addition of two electrons and the enzyme nitrate reductase:

$$HNO_3 + 2H \rightarrow HNO_2 + H_2O$$

B. The formation of an unstable intermediate, hyponitrite, with the help of nitrite reductase.

$$2NO_2 + 6H \rightarrow N_2O_2H_2 + 2H_2O$$

C. At high temperatures hyponitrite is reduced by hyponitrite reductase to nitrous oxide:

$$N_2O_2H_2 \rightarrow N_2O + H_2O$$

D. At low temperatures nitrogen gas is formed:

$$N_2O_2H_2 + 2H \rightarrow N_2 + H_2O$$

This effect of temperature on the end products of denitrification is illustrated by *Denitrobacillus* which produces approximately equal quantities of nitrogen and nitrous oxide at 30°C but 63% nitrous oxide at 51°C.

E. The reduction of hyponitrite to ammonia proceeds under microaerophilic conditions and recognizes hydrazine as an intermediate. Microorganisms implicated in this reaction include *Bacillus subtilis* and some *Azotobacter* species.

Very little is known concerning the enzyme systems of denitrification. Nitrate reductase is a collective term for one or more nitrate reducing systems each involving complex cytochrome, flavoprotein and NAD coenzyme mechanisms. It is suggested that oxygen levels above 5% interfere with the specific cytochrome functions in denitrification and ammonia formation occurs.

3.2. *Non-Biological Loss of Nitrogen*

As the process of microbial nitrate reduction becomes better understood it is increasingly evident that there are losses which cannot be explained by biological mechanisms alone. There is now evidence that chemo-denitrification of nitrate and nitrite occurs in soil and a number of possible pathways have been proposed.

A. Under acid conditions nitrite decomposes to yield nitrous oxide which is chemically oxidized to nitric oxide which, in turn, reacts with water to form nitric acid. In a fluctuating soil environment some gaseous nitrogen may be lost during this sequence of events.

B. It is possible for nitrous acid to react with amino acids and yield molecular nitrogen (Van Slyke reaction):

$$RNH_2 + HNO_2 \rightarrow ROH + H_2O + N_2$$

C. Ammonia may interact with nitrous acid to give nitrogen and water:

$$NH_3 + HNO_2 \rightarrow N_2 + 2H_2O$$

D. The volatilization of ammonia is considerable under certain conditions (especially alkaline soils) and as much as 25% of ammonium supplied in fertilizers or formed microbiologically may be lost to the atmosphere.

3.3. *Ecological Factors*

Since the transition of nitrate to nitrogen gases is a reduction process energy is required from an oxidizable substrate. The presence in soil of easily oxidized organic matter may induce denitrification by using up oxygen and providing an alternative electron acceptor. Less rapidly decomposed organic compounds (lignin, cellulose) will not stimulate denitrification to the same extent.

4. Fixation

To maintain a steady-state level of nitrogen in the biosphere, losses (through denitrification, leaching, harvesting of crops, etc.) must be compensated by gains (fertilizer application, microbial fixation etc.). Gaseous nitrogen may be used as a nitrogen source by a diverse microbial flora, converted to protein and injected into the mineralization/nitrification sequence on the death of the microorganism. This is described as nitrogen fixation.

TABLE 3.6
Nitrogen-fixing microorganisms

Asymbiotic			
Aerobes		Obligate and Facultative Anaerobes	
Heterotrophs	Autotrophs	Heterotrophs	Autotrophs
Azotobacter	*Nostoc*	*Clostridium*	*Chromatium*
Beijerinckia	*Anabena*	*Desulphovibrio*	*Chlorobium*
Derxia	*Tolypothrix*	*Klebsiella*	*Rhodospirillum*
Mycobacterium	*Gleocapsa*	*Bacillus*	
Nocardia	*Fischerella*	*Achromobacter*	

Symbiotic

Root Nodules Leguminous Plant + *Rhizobium*
Lichens Fungus + Blue-Green Algae
Some Liverworts, Cycads, Water Ferns + Blue-Green Algae
Leaf Nodules *Psychotria* + *Klebsiella*
Alnus, Myrica, Coriara + Unconfirmed Symbionts

The microorganisms reported to fix nitrogen (Table 3.6) fall into two broad categories: those which are free-living (asymbiotic) and those which form mutally beneficial relationships with plants (symbiotic).

4.1. Free-Living Nitrogen Fixers

Probably the most important asymbiotic nitrogen fixers (in terms of efficiency of introducing nitrogen into the nitrogen cycle) are the aerobic heterotroph *Azotobacter*, the anaerobic heterotroph *Clostridium* and, in aquatic environments, the photoautotrophic blue-green algae. *Azotobacter* species, fixing as much as 350 μg of nitrogen (ml of culture medium)$^{-1}$ day^{-1}, may number 1000 (gram dry weight of soil)$^{-1}$. *Clostridium*, although less efficient at nitrogen fixing, may be present in even larger number in anaerobic microsites in soils with an overall aerobic environment. Blue-green algae are important nitrogen fixers in paddy fields—the best known genera being *Anabaena, Calothrix, Nostoc, Cylindrospermum* and *Tolypothrix*.

Other important genera of bacteria with the ability to fix nitrogen include: *Beijerinkia, Derxia* and *Klebsiella* and the photosynthetic bacteria *Chromatium* and *Rhodospirillim*.

4.2. *Symbiotic Nitrogen Fixers*

The importance of legumes in agricultural practice has been recognized for many years although it wasn't until the end of the last century that the relationship between root nodule formation and the nitrogen economy of the soil was established. A large proportion of the plant family Leguminoseae is capable of forming nitrogen-fixing nodules and a number of them are economically important crops (Table 3.7). The bacteria responsible for the formation of nodules

TABLE 3.7
Nitrogen fixation by commercially important crops

Plant	kg N_2 fixed ha^{-1} $year^{-1}$
Lucerne	126–335
Red Clove	84–193
Vetch	89–158
Pea	82–148
Cowpea	64–131
Soybean	64–118

and the subsequent fixation of nitrogen belong to the genus *Rhizobium*. The bacteria/plant relationship is described as symbiotic—both components benefiting from the association. Rhizobia and legumes may live independently and it has recently been demonstrated that free-living rhizobia can still fix nitrogen.

The stages of infection and nodule formation are now well defined and are outlined in Fig. 3.17. The presence of bacteroids (Plate 3.2) and the protein leghaemoglobin are essential for the effective fixation of nitrogen. The nitrogen compounds synthesized in the nodule (primarily amino-acids) are translocated through the plant and are mineralized in the normal manner upon the death of the legume.

A variety of non-leguminous nitrogen-fixing symbionts have been reported, none of them utilizing *Rhizobium* species and none of

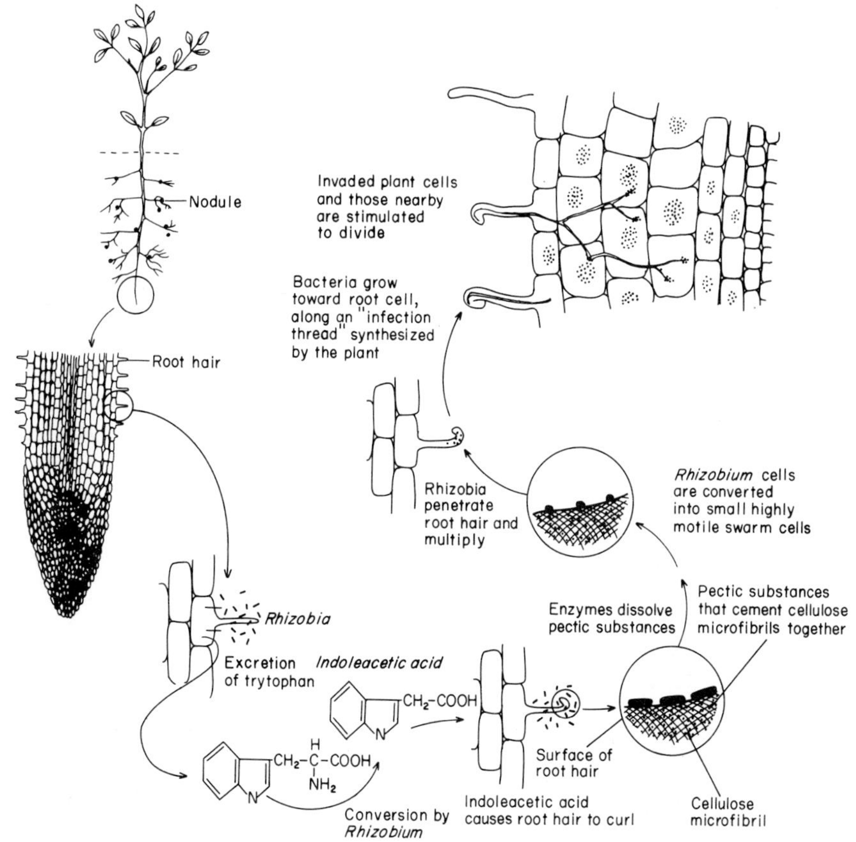

Fig. 3.17. Stages in the formation of a root nodule.
Thomas D. Brock, Biology of Microorganisms, (c) 1970, p. 409. Reprinted by permission of Prentice-Hall, Inc., Englewood Cliffs, New Jersey.

obvious agricultural importance. This is not to say that the nitrogen cycle in some environments is not influenced by such plants as the alder or bog myrtle.

4.3. Biochemistry

The first identifiable product of microbial nitrogen fixation is ammonia. The reduction of nitrogen to form ammonia involves nitrate reductase and at least two intermediates, possibly diimide and hydrazine (Fig. 3.18)

$$N_2 \xrightarrow{+2H} HN{=}NH \xrightarrow{+2H} (NH_2)_2 \xrightarrow{+2H} 2NH_3$$

Fig. 3.18. Reduction of nitrogen.

Ammonia then combines by reductive condensation with α-keto-glutaric acid to form the first organic derivative, glutamic acid (Fig. 3.19) which is converted to other amino acids by transamination.

$$NH_3 \xrightarrow[\text{α-Keto-glutaric acid}]{} \text{iminoglutaric acid} \xrightarrow[\text{glutamic dehydrogenase}]{} \text{glutamic acid}$$

Fig. 3.19. Glutamic acid synthesis.

4.4. Ecological Factors

Microorganisms will only fix atmospheric nitrogen as a last resort and therefore addition of nitrate and ammonia will effectively inhibit fixation. The presence in the soil, plant or growth media of molybdenum is essential as this is a micronutrient which activates nitrate reductase (a molybdenum-flavoprotein complex). Optimum concentrations are of the order 0.10–1.0 ppm. In addition, cobalt in small quantities (0.001 ppm) is important in the synthesis of nitrate reductase and both calcium and zinc are essential micronutrients for nitrogen fixation by *Azotobacter*.

The pH of the soil system will determine the type of micro-organism present. *Azotobacter* species are intolerant of acid environments, pH 7.2-7.6 being their optimum range. *Beijerinkia*, on the other hand, is an acid tolerant genus. *Derxia* grows well in a pH range of 5.0-8.5 whilst *Clostridium* fixes nitrogen between pH 4.0-9.0.

5. Immobilization

At any one time a proportion of nitrogen will be unavailable to the direct requirements of the nitrogen cycle because of immobilization. This subtraction of nitrogen may be ephemeral (as protein in microbial cells) or of a more permanent nature as organic nitrogen in harvested plants.

6. Nitrogen Compounds Used in Agriculture

The production of inorganic nitrogen fertilizers has increased faster than other chemical fertilizers and, more than any other single factor, has contributed to improvements in crop yields during the last twenty years. The principal nitrogenous fertilizers and their nitrogen contents are listed in Table 3.8.

TABLE 3.8
Nitrogen content of nitrogenous fertilizers

Material	Approximate Nitrogen content (%)
Anhydrous ammonia	82
Urea	37
Calcium cyanamide	35
Ammonium nitrate	35
Ammonium sulphate	21
Diammonium phosphate	21
Ammonium phosphate sulphate	16
Sodium nitrate	16
Organic products (manure, fishmeal etc.)	1–12

Ammonia is the principal source of nitrogen in fertilizers, with greater than 90% of them actually consisting of, or being manufactured from, ammonia (Fig. 3.20).

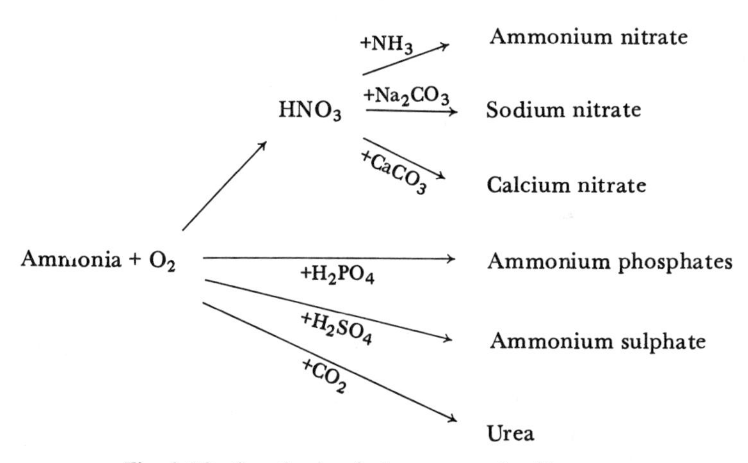

Fig. 3.20. Synthesis of nitrogenous fertilizers.

Only calcium cyanamide and Chilean nitrate of soda do not involve ammonia in their manufacture. Three of the highest nitrogen-yielding fertilizers (ammonia, urea and ammonium nitrate) are the most extensively used.

The nitrogen in most chemical fertilizers is readily soluble and rapidly nitrified so that it very soon becomes available to plants. This turnover is sometimes too rapid if the plant cannot absorb or does not need all the nitrate. Residual nitrate may then be removed from the plant root environment by leaching, rendering the fertilizer less efficient and contributing to the eutrophication of water and related problems (see p. 106). As a result, attempts have been made over the years to retard the release of nitrogen from fertilizers by the addition of nitrification inhibitors such as N-Serve(trichloromethyl pyridine) or complexing the fertilizer with a protective agent (sulphur).

Mammalian Toxicity

Nitrate itself is relatively innocuous to mammals. When reduced to nitrite, however, it readily combines with ferrous iron in haem to form ferric iron and reduce the ability of the blood to carry oxygen. This condition is known as methaemoglobinaemia and can lead to asphyxiation and death. Fatalities have been reported in Europe and America.

Nitrate enters the mammalian body, in drinking water, presumably derived from streams into which nitrate has leached. The nitrate itself may have originally come from fertilizers or human and domestic animal sewage. Levels of nitrate in excess of 45 ppm render water unsuitable for drinking. In addition some plants (notably spinach) when grown on soil high in nitrogen fertilizer, may accumulate nitrate without any subsequent convertion to protein.

The reduction of nitrate to nitrite may be mediated by airborne bacteria after exposure of cooked plant food to the atmosphere, by the intestinal microflora (especially in infants where stomach pH and intestinal disturbances encourage nitrite production) or even by enzymes within plants themselves. Problems of human toxicity can be avoided by (a) monitoring nitrate levels in drinking water, (b) not keeping prepared vegetables at room temperature (especially spinach, squash, beets and carrots), (c) using low levels of nitrogen fertilizer during growth of susceptible plant species, (d) ensuring that plant levels do not exceed 300 ppm, and (e) not feeding prepared high risk foods to infants less than six months old.

In livestock the ingestion of nitrate-containing plants may induce abortions.

In addition to the hazards described above, nitrite, in combination with organic components in food or even in the human body, forms nitroso compounds, some of which are powerful carcinogens. Nitrate in canned food may cause tin-plate corrosion.

PHOSPHORUS

Phosphorus is a component of nucleic acids, phospholipids, inositol phosphates, sugar phosphates, coenzymes, etc. Plant residues may contain up to 0.5% phosphorus and, *in toto*, phosphorus of organic origin (plants, microbes, animals) contributes up to 70% of that element in soil. Plants are, for the most part, unable to take up organic phosphorus and only utilize it in an inorganic soluble form such as orthophosphate ($H_2PO_4^-$).

Whilst there is potentially plenty of phosphorus in soils, much of it exists as insoluble inorganic complexes with iron, magnesium and aluminium, especially at acid pH. As a result, even soils with high phosphorus levels may support crops with a deficiency of phosphorus.

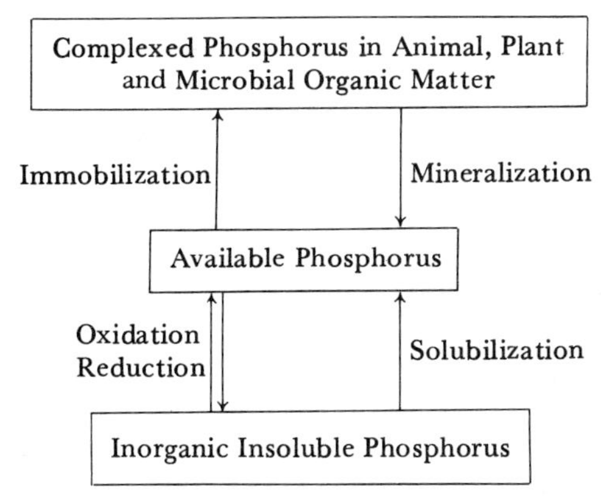

Fig. 3.21. The phosphorus cycle.

Unlike nitrogen, phosphorus does not undergo any valency changes during assimilation and enters into organic combinations as phosphate. Nor is it likely that changes occur during organic matter degradation. This lack of volatility and reactivity of phosphorus produces a somewhat limited and less spectacular biological cycle (Fig. 3.21) than that of nitrogen. The involvement of soil microorganisms in phosphorus transformations may include: (a) changing the solubility of inorganic phosphorus, (b) mineralizing organic material to release orthophosphate, (c) converting available phosphorus to microbial protoplasm (immobilization), and (d) oxidation and reduction reactions.

1. Solubilization

A great many microbes (*Pseudomonas, Mycobacterium, Micrococcus, Flavobacterium, Penicillium, Aspergillus* and others) have the potential to solubilize inorganic phosphorus compounds. They may achieve this end in two basic ways—acid production or the simulation of a reducing environment. Acid may be generated by chemolithotrophs producing nitrous and nitric acids (*Nitrosomonas, Nitrobacter*) or sulphuric acid (*Thiobacillus*), by all respiring bacteria producing carbon dioxide (carbonic acid) and those which produce organic acids as metabolic by-products. This last group of acids includes formic, acetic, propionic, lactic, glycollic, fumaric, succinic, 2-keto-gluconic and others. A good example of how acid production increases phosphate availability is shown by the three calcium phosphates–monocalcium phosphate ($Ca(H_2PO_2)_2$), dicalcium phosphate ($CaHPO_4$) and tricalcium phosphate ($Ca_3(PO_4)_2$). Of these salts monocalcium phosphate has the highest water solubility whilst tricalcium phosphate is practically insoluble. The reaction moves from right to left as pH decreases (Fig. 3.22).

$$Ca(H_2PO_4)_2 \rightleftharpoons CaHPO_4 \leftrightarrows Ca_3(PO_4)_2$$

Fig. 3.22. Phosphate availability at various pH's.

Under anaerobic conditons ferric phosphate may be reduced to ferrous phosphate with an increase in solubility. Tricalcium and aluminium phosphate are not so dramatically affected. In addition, hydrogen sulphide, produced by such bacteria as *Desulphovibrio*, reacts with ferric phosphate to form ferrous sulphide and hypophosphate.

The evolution of carbon dioxide and secretion of organic acids by plant roots may also have an important influence on phosphate availability in the rhizosphere.

2. Mineralization

2.1. Phytin

The most abundant class of phosphorus-containing organics in soil are esters such as inositol phosphate, which in itself may account for some 50% of the total organic phosphorus. The hexaphosphate of myoinositol (myo-IHP or phytic acid) [XII] is a common component of soil humus, generally as its calcium-magnesium salt-phytin. Phytin is only slowly decomposed in the soil despite the abundance of microbes capable of producing phytase. This resistance of phytin to decay is the result of at least three factors–the stability of phosphate-ester linkages, its ability to form poorly soluble complexes with iron, aluminium, calcium and magnesium salts, and its adsorption to clay surfaces. Microbes involved in phosphate ester metab-

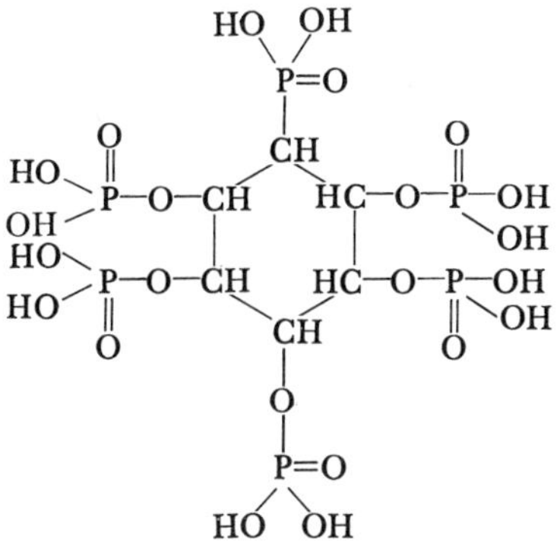

```
                HO  OH
                  \/
                  P=O
                   |
       O           CH            O
HO     ||        /    \          ||
  >P—O—CH            HC—O—P—OH
OH                  |            \
                     |              OH
HO                  |
  >P—O—CH            HC—O—P—OH
HO     ||        \    /          || \
       O           CH            O    OH
                   |
                   O
                   |
                  P=O
                  / \
                HO   OH
```

[XII]

olism include *Aspergillus, Penicillium, Rhizopus, Bacillus* and *Arthrobacter* which produce inositol and phosphate.

2.2. Nucleic Acids

In contrast to phytin, nucleic acids are rapidly dephosphorylated by microbial nucleases (see page 89).

2.3. Phospholipids

Only a small proportion of soil phosphorus is present as phospholipids or phosphatides. Among the commonest are the glycerophosphatides such as lecithin, phosphatidyl ethanolamine and phosphatidylcholine. The lecithin molecule [XIII] consists of glycerol esterified with two molecules of fatty acid and one of phosphoric acid itself esterified with choline.

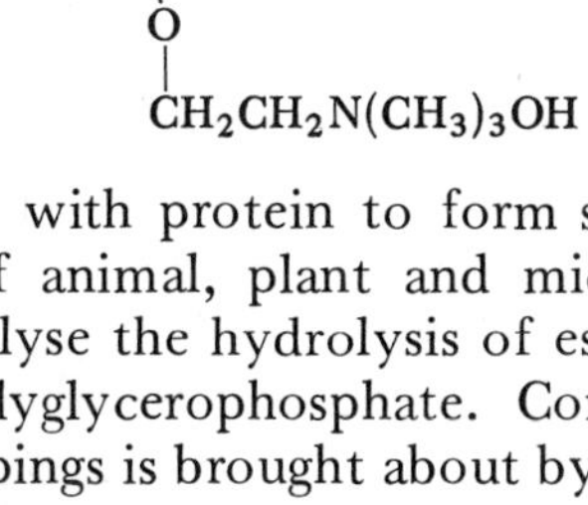

```
     O
     ||
CH2OCR
|    O
|    ||
CHOCR1
|      OH
CH2OP<
     \ O
     O
     |
     CH2CH2N(CH3)3OH
```

[XIII]

Lecithin complexes with protein to form stable lipoprotein conjugates. A number of animal, plant and microbial enzymes (lecithinase A and B) catalyse the hydrolysis of ester linkages to release fatty acids and cholyglycerophosphate. Complete hydrolyses of phosphoric ester groupings is brought about by mixtures of phosphodiesterases and phosphomonoesterases.

3. Immobilization

Microbial immobolization occurs when relatively large amounts of carbon and nitrogen are available (C:P>2-300:1, N:P = 10:1). Microbes are comparatively rich in phosphorus–fungal mycelium 0.5-1%, bacteria 1.5-2.5% dry weight–and their demand for this element must be set against the requirements of higher plants.

4. Oxidation and Reduction

Phosphorus, like nitrogen, may exist in a number of oxidation states from the −3 of phosphine (PH_3) to the +5 of orthophosphate (H_2PO_4). Microbiological intracellular oxidation of orthophosphite to orthophosphate ($HPO_3^{2-} \rightarrow HPO_4^{2-}$) has been recorded for a number of soil heterotrophs (*Pseudomonas fluorescens, Aerobacter aerogenes, Erwinia amylovora, Aspergillus niger, Penicillium notatum* and others) although there is no evidence for this being an energy yielding reaction.

Interesting to note is that nitrogen fixation by *Azotobacter vinelandii* is inhibited by orthophosphite.

The microbial reduction of phosphate in the environment has not been conclusively demonstrated. However, it has been suggested that under anaerobic conditions the reduction of phosphate to phosphite or even the gas phosphine can occur:

$$H_3PO_4 \rightarrow H_3PO_3 \rightarrow H_3PO_2 \rightarrow PH_3$$

Fig. 3.23. Anaerobic reduction of phosphate.

Clostridium butyricum and *Echerichia coli* may mediate this change which is analagous to denitrification.

5. Phosphorus Compounds Used in Agriculture

There is a chronic shortage of available phosphorus in nature which is synonymous with soil infertility in some areas. An "average" arable soil contains around 0.1% phosphorus of which only a small proportion is available to plants. The traditional use of bone products as phosphate fertilizers has been replaced in recent years by rock phosphates, treated in various ways to increase solubility and thus efficiency of plant utilization. Nevertheless, it has been sugges-ted that plant recovery of applied phosphate fertilizer is less than 25%. Examination of drainage water shows that this element, unlike nitrogen, does not show any appreciable leaching through soil. Apparently, phosphate becomes unavailable to plants due to a combination of pH-mediated insolubilization (Fig. 3.22), fixation by clays and microbial anabolism. Addition of readily decomposable organic matter may release, through chelation, some of the fixed phosphorus by producing organic ions (citrates, tartrates, acetates, etc.) which have a stronger affinity for calcium, iron and aluminium.

Rock phosphate may be treated with either sulphuric acid, to produce superphosphate or phosphoric acid to produce triple superphosphate. Other materials used in supplying fertilizer phosphorus are diammonium phosphate (53% P_2O_5 and 21% N), monoammonium phosphate (48 P_2O_5 and 11% N), ammonium phosphate sulphate (20% P_2O_5 and 16% N) and basic slag (10% P_2O_5). In addition, the use of nitric acid to make nitrophosphates has recently increased, due to the prohibitive cost of sulphuric acid.

OTHER FERTILIZERS

1. Potassium

In addition to nitrogen and phosphorus, potassium is an essential macronutrient of growing plants. Its primary sources in soil include feldspars, micas and the clay mineral illite from which potassium ions are released by weathering. These ions may then be leached from the soil or become adsorbed onto clay colloids. The ability of plants to extract potassium from soil varies considerably from species to species.

Potassium does not undergo any transformations in living tissue but exists in solution where it functions in a variety of metabolic reactions. There is little, if any information concerning any direct microbial transformation of potassium. Soil deficiencies of potassium can be alleviated with fertilizers or manure.

2. Sulphur

Sulphur shortages in soil are rare although reports of deficiencies in legumes and tea plants have occurred. Plants absorb sulphate ions which have either been supplied by the weathering of mineral sulphides or the oxidation of sulphur-containing organic matter. A variety of microbial transformations of sulphur and the associated pollution problems are discussed elsewhere in this volume (Chapter 8).

Many sulphur-deficient soils are also phosphorus deficient and dressing with superphosphate (8% P, 12% S) will alleviate both problems. If only sulphur is required, calcium sulphate or finely divided elemental sulphur will suffice. More recently, concentrated liquid fertilizers—such as ammonium polysulphide and thionates—have been developed.

Other occasional mineral additions to soil include manganese, iron, molybdenum, selenium, silica, copper and zinc.

EUTROPHICATION

Man's dependence upon an intensive agricultural system brings about a conflict between resource exploitation and pollution control. This is due to the incompatibility between maintaining a monoculture whilst ensuring that surface waters remain unaffected by fertilizer and sewage leachate and run-off.

An equilibrium between photosynthesis and respiration is a prerequisite for the constant chemical and biological composition of water. In other words, the rate of cell production must at least approximate to the rate of cell decay. If photosynthesis suddenly becomes significantly greater than respiration there is a progressive accumulation of algae; if respiration is greater than cell production the BOD may exceed supply and anaerobic conditions develop. Departures from the $P = R$ steady-state situation occur when waters receive excess organic nutrients (stimulation of bacterial activity; $R > P$) or excess inorganic nutrients (stimulation of algal growth; $P > R$).

Lakes are, of course, dynamic systems evolving naturally (albeit slowly) as sediments accumulate or as nutrient levels of inflow water change. Human interference can accelerate this change by introducing sewage or agricultural run-off. As a result a "clean" *oligotrophic* lake can become, in a matter of a few years, a stagnant, anaerobic *eutrophic* lake. Waters in a transition between the two extremes (see Table 3.9) may be considered preferable because they support a full and varied flora and fauna.

1. Causes

It is apparent that phosphate and nitrate are key limiting nutrients in oligotrophic lakes and that their introduction will often initiate eutrophication. It is also suggested that carbon (as carbon dioxide) may become limiting in a well-developed eutrophic lake.

TABLE 3.9
Comparison of the properties of oligotrophic and eutrophic waters

Property	Eutrophic waters	Oligotrophic waters
Appearance	Fairly clear, green, low light penetration.	Very clear. High light penetration.
Hardness	Often hard.	Usually soft.
Odour and taste	Often, but not always foul.	None or "peaty".
Fish	None or coarse.	Present-often salmon and trout.
Oxygen content	Low, variable with season and depth.	Near saturation.
Treatment for water supplies	Slow filtering, may block sand filters.	Easily filtered.

Most phosphate which enters water comes from human and agricultural sewage and industrial effluents. The contribution of detergents to the total is discussed elsewhere (Chapter 5). The leaching of phosphate fertilizers is probably insignificant in most areas although surface run-off, after sudden and high rainfall, may transport phosphates to water courses. As little as 0.01 ppm phosphorus will initiate eutrophication if other nutrients are in excess. However, as "background" soluble phosphate levels are often in excess of this (0.5 ppm), nitrogen is probably the major factor, at least in the early stages of eutrophication (N:P = 15:1).

The principal forms of nitrogen in lake water are NH_4^+, NO_3^-, NO_2^- and organic nitrogen. The relative proportions of these nutrients will depend, not only on input but also on the biological transformations described previously (p. 87). Deep lakes exhibit a dichotomic distribution-maximum concentration of nitrate at intermediate depths, denitrification at the bottom and immobilization on the surface. Concentrations of less than 0.3 ppm of nitrogen will induce eutrophication whilst levels of nitrate alone have been reported as exceeding 10 ppm. Unlike phosphorus, the leaching of fertilizer nitrate contributes significantly to this total, as does human and domestic animal sewage.

2. Microbiology

Dramatic responses by algae to nutrient input include the growth of both procaryotic (*Monodus*) and eucaryotic (*Cladophora*) species.

These plants may remove carbon dioxide at such a rate that carbonate levels decrease and the lake becomes strongly alkaline ($pH > 8$). This, in turn, encourages the development of blue-green algae (*Oscillatoria, Anabaena*) which have the additional competitive advantage of being able to fix nitrogen if that nutrient becomes limiting. The success of the blue-greens at this stage in the development of an eutrophic lake is at the expense of other phytoplankton. Thus the potential variety of food sources for herbivores is restricted and certain fishes may be eliminated from the ecosystem. During the initial surge of algal growth, grazing crustaceae (*Daphnia*) will often retard the rate of eutrophication.

Once a certain density of algal growth has been reached carbon and light dioxides become limiting factors and the death of phytoplankton rapid. This input of organic matter into the mineralization cycles of the lake puts an impossible demand on the oxygen, depleting the water to such an extent that fish cannot survive even though there is an almost limitless supply of food. The very slow (now anaerobic) decay of plant material provides for a gradual filling-in of the lake, is odorous (ammonia, hydrogen, sulphide, amines) and unsightly. Aerobic and facultative anaerobic Gram negative rods predominate in lakes undergoing eutrophication and include members of the genera *Pseudomonas, Achromobacter, Alcaligenes* and *Flavobacterium.*

Recommended Reading

Benemann, J. R. and Valentine, R. C. (1972). The pathways of nitrogen fixation. *Advances in Microbial Physiology* 8, 59.

Commoner, B. (1970). Threats to the integrity of the nitrogen cycle: nitrogen compounds in soil, water, atmosphere and precipitation. *In* "Global Effects of Environmental Pollution" (S. F. Singer, Ed.), D. Reidel Publishing Co.

Cosgrove, D. J. (1967). Metabolism of organic phosphates in soil. Ch. 9. *In* "Soil Biochemistry" (A. D. McLaren and G. H. Peterson, Eds.), Vol. 1, Marcel Dekker, New York.

Dugan, P. R. (1972). "Biochemical Ecology of Water Pollution." Part II. Plenum.

Halstead, R. L. and McKercher, R. B. (1975). Biochemistry and cycling of phosphorus. Ch. 2. *In* "Soil Biochemistry" (E. A. Paul and A. D. McLaren, Eds.), Vol. 4, Marcel Dekker, New York.

Hawkes, H. A. (1965). The ecology of sewage bacteria beds. In "Ecology and the Industrial Society" (G. T. Goodwin, R. W. Edwards and J. M. Lambert, Eds.), Blackwell.

Jurasek, L. Colvin, J. R. and Whitaker, D. R. (1967). Microbiological aspects of the formation and degradation of cellulosic fibres. *Advances in Applied Microbiology* 9, 131.

Keeney, D. R., Herbert, R. A. and Holding, A. J. (1971). Microbiological aspects of pollution of fresh water with inorganic nutrients. *In* "Microbial Aspects of Pollution" (G. Sykes and F. A. Skinner, Eds.), Academic Press.

Kirsch, E. J. and Sykes, R. M. (1971). Anaerobic digestion in biological waste treatment. *Progress in Industrial Microbiology* 9, 155.

Oglesby, R. T., Christman, R. F. and Driver, C. H. (1967). The biotransformation of lignin to humus-facts and postulates. *Advances in Applied Microbiology* 9, 171

Painter, H. A. (1971). Chemical, physical and biological characteristics of wastes and waste effluents. *In* "Water and Water Pollution Handbook" (L. L. Ciaccio, Ed.), Vol. 1, Marcel Dekker, New York.

Pike, E. B. and Curds, C. R. (1971). The microbial ecology of the activated sludge process. *In* "Microbial Aspects of Pollution" (G. Sykes and F. A. Skinner, Eds.), Academic Press.

Timell, T. E. (1964 and 1965). Wood hemicelluloses. *Advances in Carbohydrate Chemistry* 19 and 20.

4 Hydrocarbons

INTRODUCTION

The term hydrocarbon embraces all those organic substances solely composed of carbon and hydrogen. Most living organisms either contain or produce small amounts of hydrocarbons. For example, many fermentative bacteria in soil, sewage plants and the ruminant stomach obtain their energy for growth by oxidizing fatty acids and simple alcohols, the resulting electrons being used to reduce carbon dioxide to the simplest of the hydrocarbons, methane (see Chapter 3). Some of this methane is subsequently oxidized by specialized bacteria capable of utilizing one-carbon compounds although most is released into the atmosphere.

In plants, high molecular weight hydrocarbons (C_{25}-C_{33}) are major components of the surface waxes of leaves. In addition, higher plants, algae and photosynthetic bacteria, synthesize cartotenoids (e.g. β-carotene) which are unsaturated hydrocarbons functioning as accessory pigments in photosynthetic systems. Some plants produce unsaturated terpenoid hydrocarbons such as limonene and α-phellandrene.

Hydrocarbons are also synthesized by animals and may occur as components of insect cuticular lipids. Those, for example, of the common cockroach, *Periplenata americana*, are largely composed of hydrocarbons (75-85%), primarily C_{25}-C_{29} *n*-alkanes, a C_{26} branched alkane and C_{27}, C_{41} and C_{43} alkenes. Mammalian skin secretions also contain hydrocarbons and in most microorganisms long chain alkanes account for about one percent of the total lipids.

There are extensive deposits of complex mixtures of hydrocarbons on the earth, both below ground (reserves at least 74.5×10^9 tonnes) and on the surface (shale oil, estimated 66×10^9 tonnes). These are thought to have been produced originally by the combined effects of heat and pressure on organic material. It is due to Man's efforts in the last two hundred years that large amounts of crude oil have been released from below the earth's surface to become widely distributed in the biosphere. This has occurred as a result of the need for oil as a

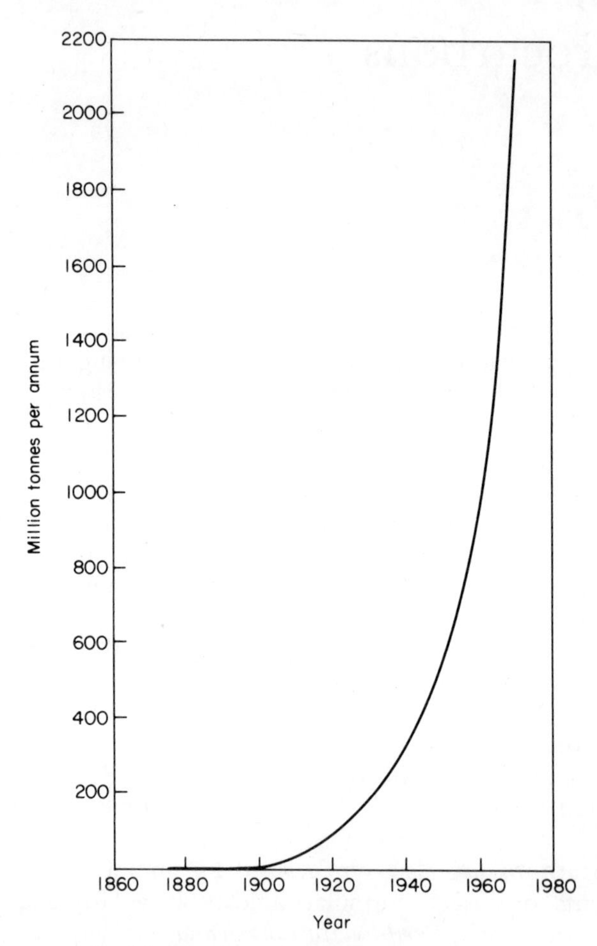

Fig. 4.1. World crude oil consumption.

fuel, a lubricant and the main raw material of the petrochemical industry. The world consumption of petroleum had risen from 20 million tonnes in 1900 to 2145 million tonnes in 1969, (Fig. 4.1) and, although most of the components are not in themselves alien to biological systems, the massive quantitites that are now released certainly are. Crude oil is refined to yield a variety of fractions with widely different uses. The gases (methane, ethane, propane, butane, acetylene, ethylene, propene) are used both as fuels and as starting materials for chemical syntheses. The low boiling point liquid hydrocarbons (petrol, paraffins) function primarily as propellants in internal combustion engines, the higher boiling point fractions as lubricants and the aromatic fractions, such as benzene, toluene and some of the alicyclic components (e.g. cyclohexane and methylcyclohexane) as important precursors in the synthesis of petrochemicals, detergents (Chapter 5) and polymers such as Nylon (Chapter 6). The high molecular weight solid waxes have a variety of uses, particularly as polishes.

The most blatant, and all too frequently detrimental environmental effect of hydrocarbons arises from large-scale local release, such as occurs when an oil tanker is damaged or there is an uncontrolled leakage from offshore oil wells. Major oil spills on land itself are rare and oil pollution problems are generally associated with water pollution. The long-term effects of oil pollution are still poorly understood because major contamination has occurred only during the last 20-25 years and more dramatically during the last ten years with the introduction of the supertanker and the development of off-shore oil fields.

As a result of the "energy crisis" and the associated increase in the price of crude oil, there is renewed interest in recovering oil from shale and tar sands. These have very different compositions from crude oil, containing mainly complex alicyclic hydrocarbons which are relatively resistant to microbial attack. Since microbes play an important role in the degradation of hydrocarbons in the environment, great care in the handling of this new oil will be necessary if serious pollution problems are to be avoided.

CHEMISTRY

Hydrocarbons include a vast array of carbon-hydrogen compounds, whose physical and chemical properties are determined, both by the numbers of atoms and their arrangement in the molecule. Molecular weights range from as low as sixteen (methane) to approximately 1000, encompassing *en route* a gamut of physical properties including compounds that are gases, those that are volatile liquids, those that are high boiling point liquids and some which are solids. The various classes of hydrocarbons are discussed below.

1. Alkanes

The alkanes (paraffin hydrocarbons) have the general formula, C_nH_{2n+2}. Compounds with linear carbon chains are known as *n*-alkanes (Fig. 4.2), but for compounds with more than two carbon

CH_4 *methane* $CH_3{-}CH_3$ *ethane* $CH_3{-}CH_2{-}CH_2{-}CH_3$ *n-butane* $CH_3{-}(CH_2)_6{-}CH_3$ *n-octane*

$CH_3{-}(CH_2)_{14}{-}CH_3$ *n-hexadecane* $CH_3{-}(CH_2)_{68}{-}CH_3$ *n-heptacontane*

Fig. 4.2. Examples of *n*-alkanes.

atoms there is a possibility of structural isomers (branched chain alkanes, Fig. 4.3). There are eighteen possible structural isomers of the C_8 alkanes and the figure for the C_{40} alkanes is about 6×10^{13}.

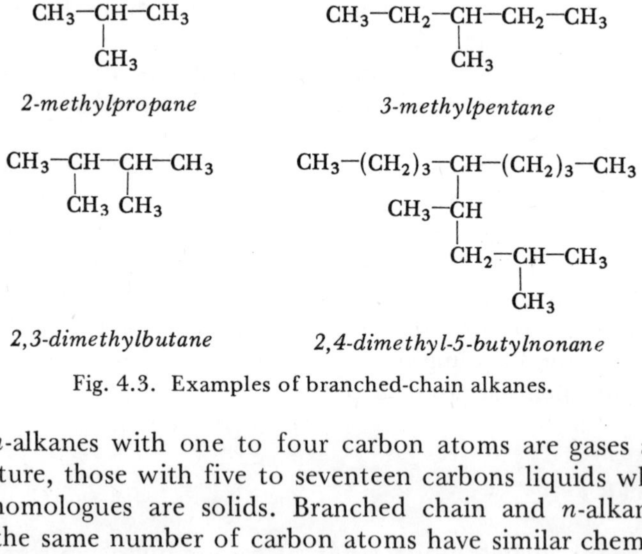

$CH_3{-}CH(CH_3){-}CH_3$ *2-methylpropane*

$CH_3{-}CH_2{-}CH(CH_3){-}CH_2{-}CH_3$ *3-methylpentane*

$CH_3{-}CH(CH_3){-}CH(CH_3){-}CH_3$ *2,3-dimethylbutane*

$CH_3{-}(CH_2)_3{-}CH[CH(CH_3){-}CH_2{-}CH(CH_3){-}CH_3]{-}(CH_2)_3{-}CH_3$ *2,4-dimethyl-5-butylnonane*

Fig. 4.3. Examples of branched-chain alkanes.

The *n*-alkanes with one to four carbon atoms are gases at room temperature, those with five to seventeen carbons liquids whilst the higher homologues are solids. Branched chain and *n*-alkanes containing the same number of carbon atoms have similar chemical and physical properties. All are characterized by being chemically inert and the few reactions that they will undergo usually require high temperatures or catalysts. Hexane, for example, can be brominated in the presence of light (Fig. 4.4).

$$C_6H_{14} + Br_2 \xrightarrow{\text{light}} C_6H_{13}Br + HBr$$

Fig. 4.4. Bromination of hexane.

When heated to 500-700°C, higher alkanes "crack" to form mixtures of short chain saturated and unsaturated hydrocarbons. This process is important in the petrochemical industry because a greater proportion of low molecular weight hydrocarbons are required than are supplied by the crude oil itself. Perhaps the most characteristic reaction of alkanes is burning in the presence of oxygen.

2. Alkenes

The physical properties of the alkenes are similar to those of the corresponding alkanes, but their unsaturated nature makes them much more chemically reactive and, as a result, important synthetic intermediates in the petrochemical industry (Fig. 4.5).

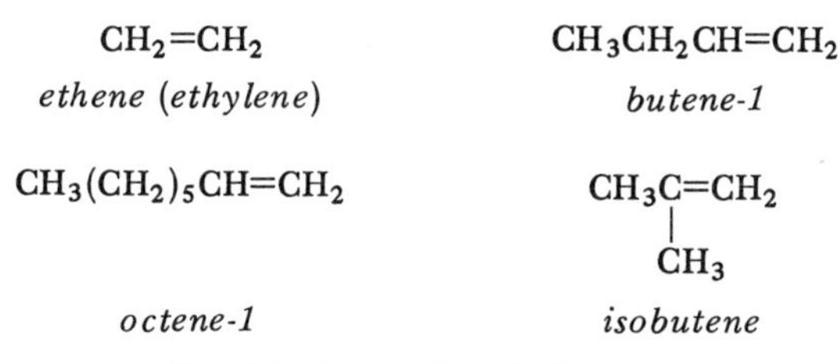

Fig. 4.5. Examples of alkenes.

Hydrocarbons with two double bonds separated by one or more saturated carbon atoms have very similar chemical properties to those alkene analogues with one double bond. However, alkenes with conjugated double bonds behave rather differently due to resonance stabilization. The simplest of these compounds is butadiene ($CH_2{=}CH{-}CH{=}CH_2$), which is highly reactive and can be readily polymerized. This compound, which is either made from petroleum by "cracking" or by the dehydrogenation of butane, is of great importance in the petrochemical industry. Present world production is about three million tonnes, of which 90% is used to make synthetic elastomers such as styrene-butadiene rubber (Ch. 6).

3. Alkynes

The alkynes (Fig. 4.6) are chemically unique amongst the hydrocarbons because the lone hydrogen attached to the triple bonded carbon is acidic and can be replaced by copper, silver, sodium and other ions to form salts. Acetylene used to be an important industrial

$HC{\equiv}CH$ — *acetylene*

$CH_3C{\equiv}CH$ — *methylacetylene*

$CH_3CH_2C{\equiv}CCH_3$ — *pentyne-2*

$CH{\equiv}C(CH_2)_5C{\equiv}CH$ — *1,8-nonadiyne*

Fig. 4.6. Examples of alkynes.

chemical but it is relatively expensive and has been largely replaced by ethene.

4. Alicyclics

Alicyclic hydrocarbons (cycloalkanes, cycloalkenes) have similar properties to their open chain alkane analogues. However, their boiling points are some 10-20°C higher and their densities 20% greater (Fig. 4.7).

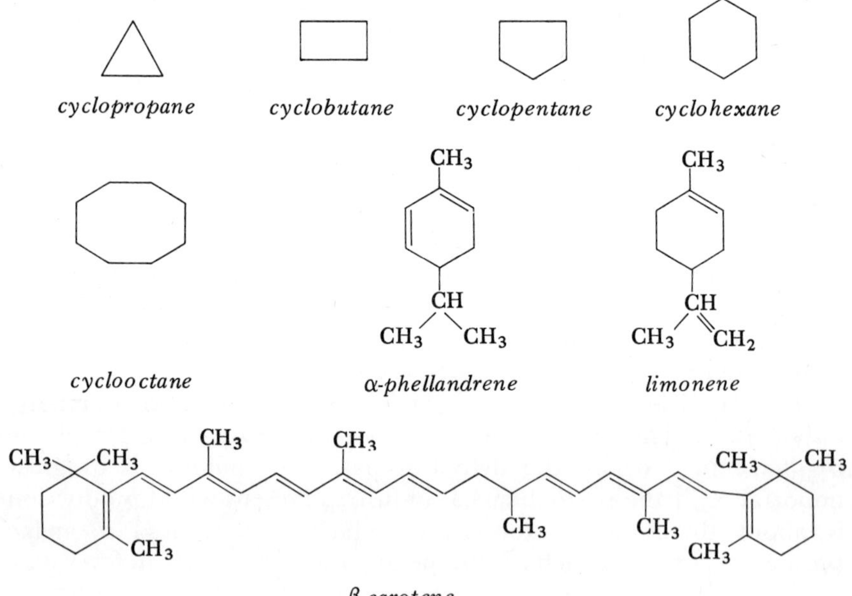

Fig. 4.7. Examples of cycloalkanes and cycloalkenes.

Alicyclic hydrocarbons are major components of crude oil and tar sands. Cyclohexane is an important petrochemical as a precursor of Nylon whilst α-phellandrene, limonene and β-carotene are alicyclic, terpenoid hydrocarbons that are common plant products.

5. Aromatics

The formulae of some common aromatic hydrocarbons are shown in Fig. 4.8.

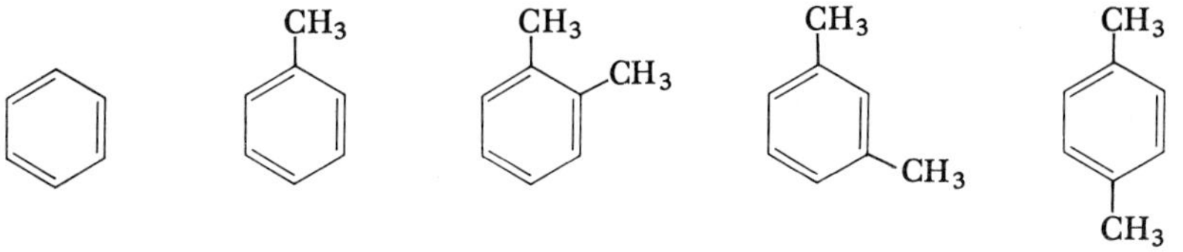

CH_3 CH_3 CH_3 CH_3 CH_3 CH_3 CH_3

benzene *toluene* *o-xylene* *m-xylene* *p-xylene*

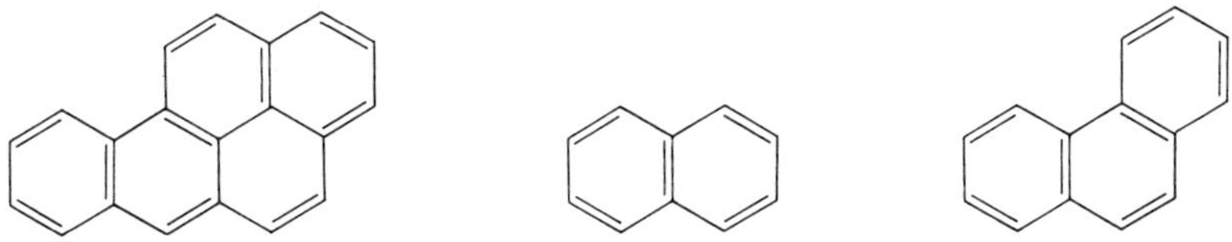

3,4-benzpyrene *naphthalene* *phenanthrene*

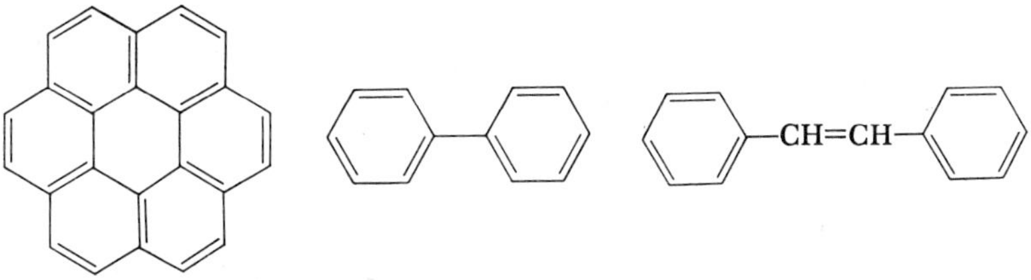

CH=CH

coronene *diphenyl* *stilbene*

Fig. 4.8. Examples of aromatic hydrocarbons.

Aromatic hydrocarbons obtained from petroleum and coal are liquids or solids at room temperature, toxic and, in some cases, carcinogenic to mammals. They are of major importance in the production of polymers, dyes, drugs and the alkylbenzene sulphonate surfactants. Table 4.1 shows the amounts of the most important aromatic hydrocarbons used in the U.S.A., Western Europe and Japan and the major uses of one of them, benzene.

6. Petroleum

Petroleum is a complex mixture of hydrocarbons combined with small amounts of other organic materials. Commercially, it is separated into fractions (Table 4.2) of different boiling ranges by distillation (refining).

TABLE 4.1

Aromatic hydrocarbon utilization (1968)

(a) Production (million tonnes)

Compound	U.S.A.	Western Europe	Japan
benzene	3.2	1.3	0.7
toluene	2.3	0.6	0.3
xylenes	1.6	0.5	0.3

(b) Benzene utilization

Used for production of:	Benzene consumption (%)	Ultimate product
ethylbenzene	35	polystrene, synthetic rubbers
phenol	25	phenolic resins
cyclohexane	17	polyamide fibres
nitrobenzene	5	alkylbenzene sulphonate surfactants, dyes, drugs.
chlorobenzene	5	
others, including alkylbenzenes	13	

Low molecular weight fractions are gaseous, known as natural gasoline and primarily consist of methane with small amounts of ethane, propane, butane and isobutane. Furthermore, trace amounts of low molecular weight alkane vapours are present.

The composition of crude petroleum is not only complex but shows considerable variation even in samples from the same locality. Typically, 200-300 different hydrocarbons are present including *n*-alkanes, branched chain alkanes, alicyclics and aromatics. The proportion of ring compounds (alicyclic and aromatic) is extremely variable (2-40% of the total). Petroleum also contains a variety of non-hydrocarbon components (Fig. 4.9) such as cycloparaffin

TABLE 4.2

Petroleum refining fractions

Fraction	Boiling point (°C)
Petroleum ether (pentanes, hexanes)	20–60
Light naphtha (hexanes, heptanes)	60–120
Gasoline (approximately 50% alkanes)	40–205
Kerosine	175–325
Gas oil	275
Residue (lubricating oil, asphalt)	>300

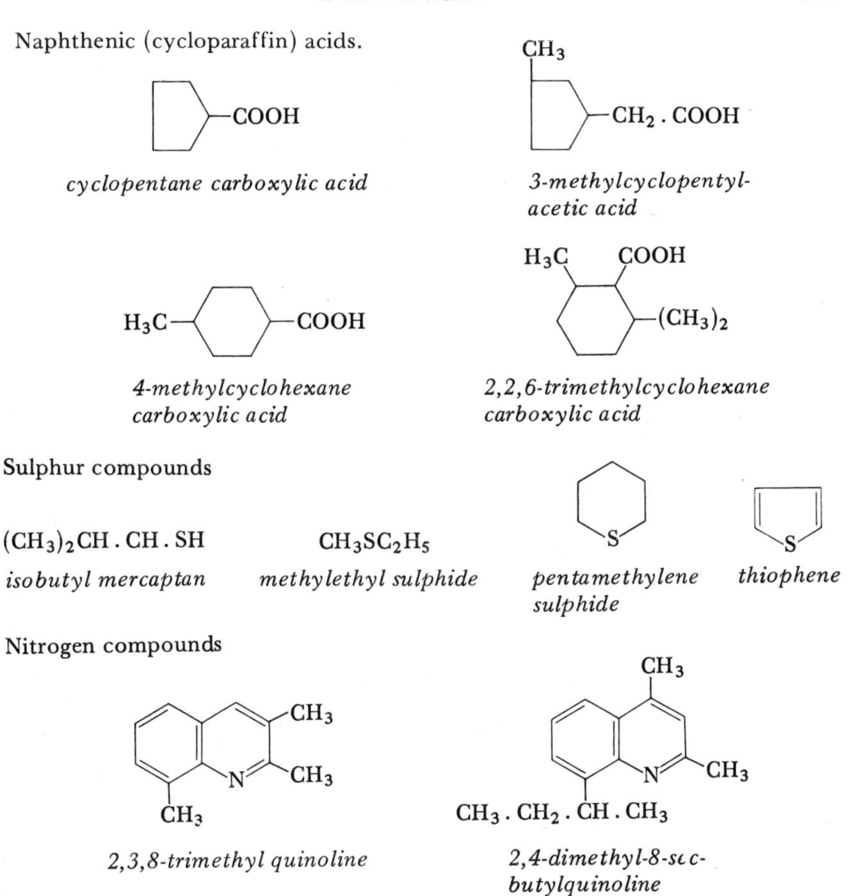

Fig. 4.9. Non-hydrocarbon components of petroleum.

(naphthenic) acids which account for 0.1-0.3% of the oil, depending on the source, and about 1% of sulphur and 0.01% of nitrogen-containing compounds.

ANALYTICAL METHODS

Hydrocarbons are only slightly water-soluble and what solubility they have decreases with an increase in molecular weight. For example, a saturated solution of hexane (M.Wt = 80) in water at 25°C has a concentration of 1.5×10^{-3} molar (138 ppm) whilst one of tetradecane (M.Wt = 198) has a concentration of 9.8×10^{-10} molar (2.0×10^{-6} ppm). This low solubility favours accurate analysis because extraction of samples with a suitable organic solvent removes the hydrocarbon from interfering water-soluble substances; the only other common solutes of organic solvents being lipids. Hydrocarbons are not very reactive chemically except with oxygen and so there are no specific chemical methods for their estimation and identification. Alkenes can undergo addition reactions, particularly halogenation and although these form the basis of tests for double bonds they are not specific to alkenes.

1. Chromatographic Techniques

Thin layer, and particularly gas-liquid chromatography are of great value in oil analysis. Examples of the use of the latter procedure are shown in Fig. 4.10. Fig. 4.10a shows the pattern obtained for an oil spill at sea after weathering for two months and the patterns obtained for extracts of oysters and scallops from the same area. Clearly the pattern in the shellfish reflects that in the oil slick. Fig. 4.10b illustrates the degradation of gas oil on incubating with marine microorganisms for two and five days.

For the identification of hydrocarbon components, combined gas-liquid chromatography mass-spectrometry is a powerful technique (see Chapter 2).

2. Spectroscopic Techniques

Unsaturated hydrocarbons (alkenes, alkynes and aromatics) absorb in the ultra-violet region of the spectrum, and consequently these wavelengths can be used to follow their fate. UV spectroscopy is not, however, suitable for saturated hydrocarbons which fail to absorb in the UV.

Infra-red analysis (wavelengths 3.3-3.5 μ) has also been used in hydrocarbon degradation studies and is a valuable tool for measuring low concentrations of oil in water, any absorption being due to the CH_3-, CH_2- and CH- groups. It is most sensitive to saturated hydrocarbons and considerably less so for unsubstituted aromatic components.

3. Miscellaneous Techniques

3.1. Observation

Since hydrocarbons are poorly soluble in water, their fate can be followed by visual means (the presence or absence of surface films) or even by loss of odour.

(a) Accumulation of hydrocarbon components of an oil slick in shellfish after two months.

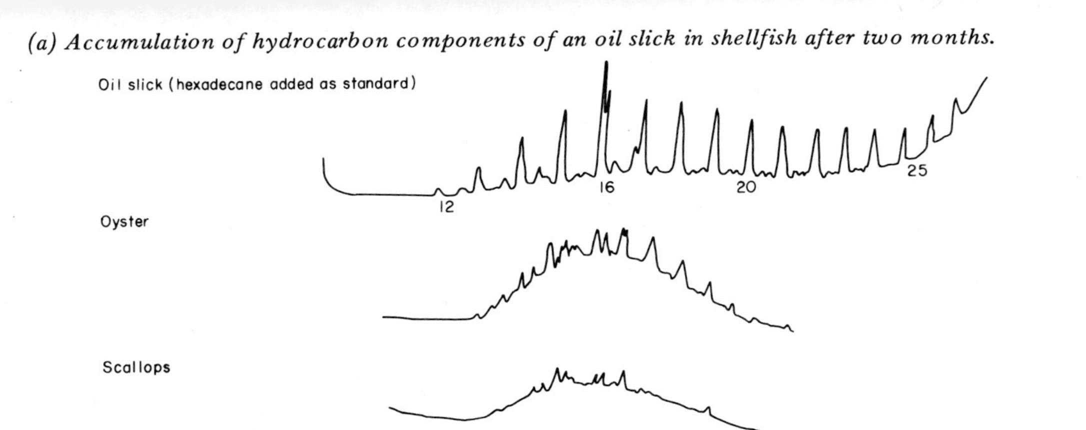

From Blumer, Souza and Sass (1970), *Marine Biol.* 5, 195.

(b) Degradation of gas-oil by a seawater microbial culture.

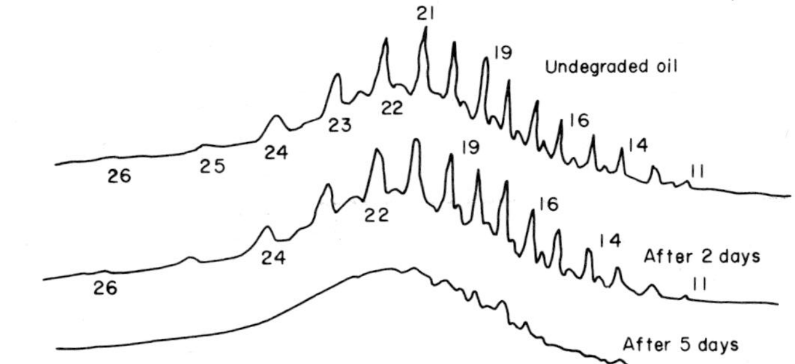

From lePetit and Barthelemy (1968), *Ann. Inst. Pasteur* 114, 149.

Fig. 4.10. Use of gas chromatography in hydrocarbon pollution studies. Numbers refer to hydrocarbon chain length. The alkane and alkene components are shown as peaks against a background of unresolved alicyclics and aromatics.

3.2. Weight Loss

Hydrocarbons can be separated readily from water or soil by extraction with organic solvents, and so it is a simple matter to weigh the amount of hydrocarbon extracted from a measured amount of sample.

3.3. Oxygen Utilization

Chemical oxygen demand (COD), combustion, biochemical oxygen demand (BOD) and respirometric measurements are commonly used in hydrocarbon metabolism studies (see Chapters 3 and 5).

HYDROCARBON DECAY

1. Non-Biological

Many hydrocarbons, including the components of petroleum, undergo complex autooxidation processes under aerobic conditions. Free-radical chain reactions occur which lead to the formation of hydroperoxides which are, for the most part, unstable and rapidly oxidized. Some hydroperoxides, in contrast, are quite stable (Fig. 4.11).

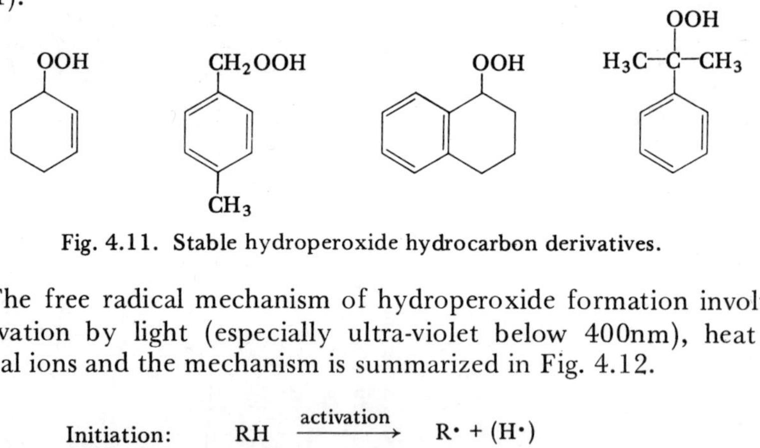

OOH　　CH_2OOH / CH_3　　OOH　　OOH / $H_3C—C—CH_3$

Fig. 4.11. Stable hydroperoxide hydrocarbon derivatives.

The free radical mechanism of hydroperoxide formation involves activation by light (especially ultra-violet below 400nm), heat or metal ions and the mechanism is summarized in Fig. 4.12.

Initiation: $RH \xrightarrow{\text{activation}} R\cdot + (H\cdot)$

Propagation: $R\cdot + O_2 \rightarrow RO_2\cdot$

$RO_2\cdot + RH \rightarrow RO_2 + R\cdot$

Termination: $R\cdot + R\cdot \rightarrow RH \quad + R(-H)$

$RO_2\cdot + \cdot OH \rightarrow ROH + O_2$

$R\cdot + R\cdot \rightarrow R{-}R$

Fig. 4.12. Free radical mechanism of hydroperoxide formation.

Some positions in open chain alkenes are not vulnerable to autooxidative attack (Fig. 4.13). At normal temperatures the tertiary –CH groups in alkanes are oxidized most readily and the primary –CH groups least readily. Alkenes are more easily attacked than

```
—C*—C=C—C—              —C—C*—C=C
    |
    C

—C*—C=C—C—C—            —C—C*—C=C—C—C
    |                         |
    C                         C
```

Fig. 4.13. Positions in open-chain alkenes resistant* to autooxidative attack.

alkanes, but not at the double bond. Various substances inhibit the rate of autooxidation (phenolic and some heterocyclic components of petroleum) whilst others stimulate it (metal ions and organometallic compounds). This latter reaction is especially important in seawater which contains metallic complexes.

A large proportion (40-60%) of crude oil spills evaporates within a few days in temperate and tropical climates, especially those components with boiling points below 370°C. Light, refined fractions evaporate completely in this time.

2. Microbial

It has been known for many years that microorganisms can degrade hydrocarbons. The earliest account of this phenomenon was published in 1895, when the degradation of a thin layer of paraffin by the fungus *Botrytis cinerea* was described. Since then the metabolism of hydrocarbons has been extensively studied and it is evident that about 20% of all microbial species examined have some capacity to degrade hydrocarbons. The most adept in this respect appear to be the soil molds, such as *Penicillium glaucum*, some yeasts (e.g. *Candida utilis*) and amongst the bacteria, species of *Pseudomonas, Nocardia* and *Mycobacteria*. Microbes have been shown to metabolize both straight and branched chain alkanes including methane upwards, alkenes, aromatics and alicyclics. The *n*-alkanes are the most readily degraded by the widest range of microbes whilst the branched chain compounds are less easily metabolized, the rate being inversely proportional to the degree of branching. The number of microorganisms that degrade aromatic hydrocarbons seems to be rather limited, but some strains of *Pseudomonas* and *Nocardia* grow readily on these compounds. Alicyclic hydrocarbons are the least susceptible to microbial attack and indeed in many cases are quite recalcitrant.

2.1. Methane

Over one hundred strains of bacteria have been isolated that can grow on methane which they are thought to oxidize to carbon dioxide by the route shown in Fig. 4.14a. These bacteria are unusual in that all

(a) route of methane oxidation

$$\underset{\text{methane}}{CH_4} \longrightarrow \underset{\text{methanol}}{CH_3OH} \longrightarrow \underset{\text{formaldehyde}}{H.CHO} \longrightarrow \underset{\text{formate}}{H.COOH} \longrightarrow CO_2$$

(b) probable mechanism of initial oxidative attack on methane by *Methylosinus trichosporium*

$$CH_4 + 2\text{Cytochrome-}c_{co}Fe^{2+} + 2H^+ + O_2 \rightarrow CH_3OH + 2\text{Cytochrome-}c_{co}Fe^{3+} + H_2O$$

N.B. Cytochrome-c_{co} is an unusual carbon monoxide-binding *c*-type cytochrome.

Fig. 4.14. Microbial metabolism of methane.

but one strain will only grow on methane, methanol or dimethylether. Some will co-oxidize (Chapter 2) the higher homologues, ethane, propane and butane, whilst growing on methane to form the corresponding acids and ketones. For example, propane is converted to acetone and propionic acid. The conversion of methane to methanol involves a complex monooxygenase system (Fig. 4.14b) whilst further oxidation to carbon dioxide involves two or possibly three dehydrogenases. Carbon is incorporated into cell material at the level of formaldehyde, either by a pentose pathway (Fig. 4.15a) the serine pathway (Fig. 4.15b) or both depending on the species.

2.2. Higher Alkanes

The routes and detailed enzymology of higher alkane metabolism have been studied extensively in recent years and it is clear that several mechanisms operate in different microorganisms. There appears to be three basic modes of initial attack on the alkane molecule (Fig. 4.16).

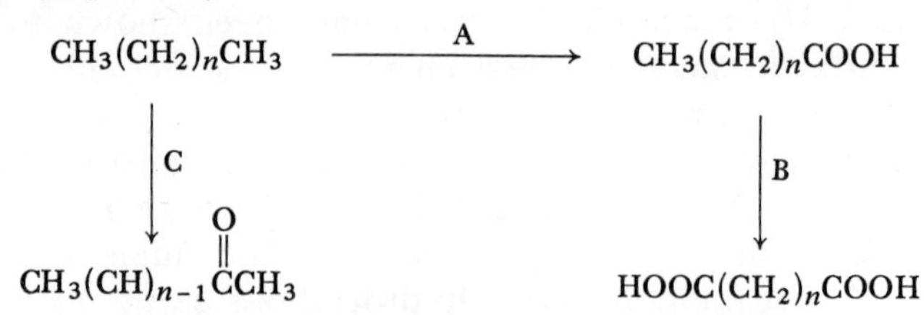

Fig. 4.16. Initial enzymic attack on alkanes.

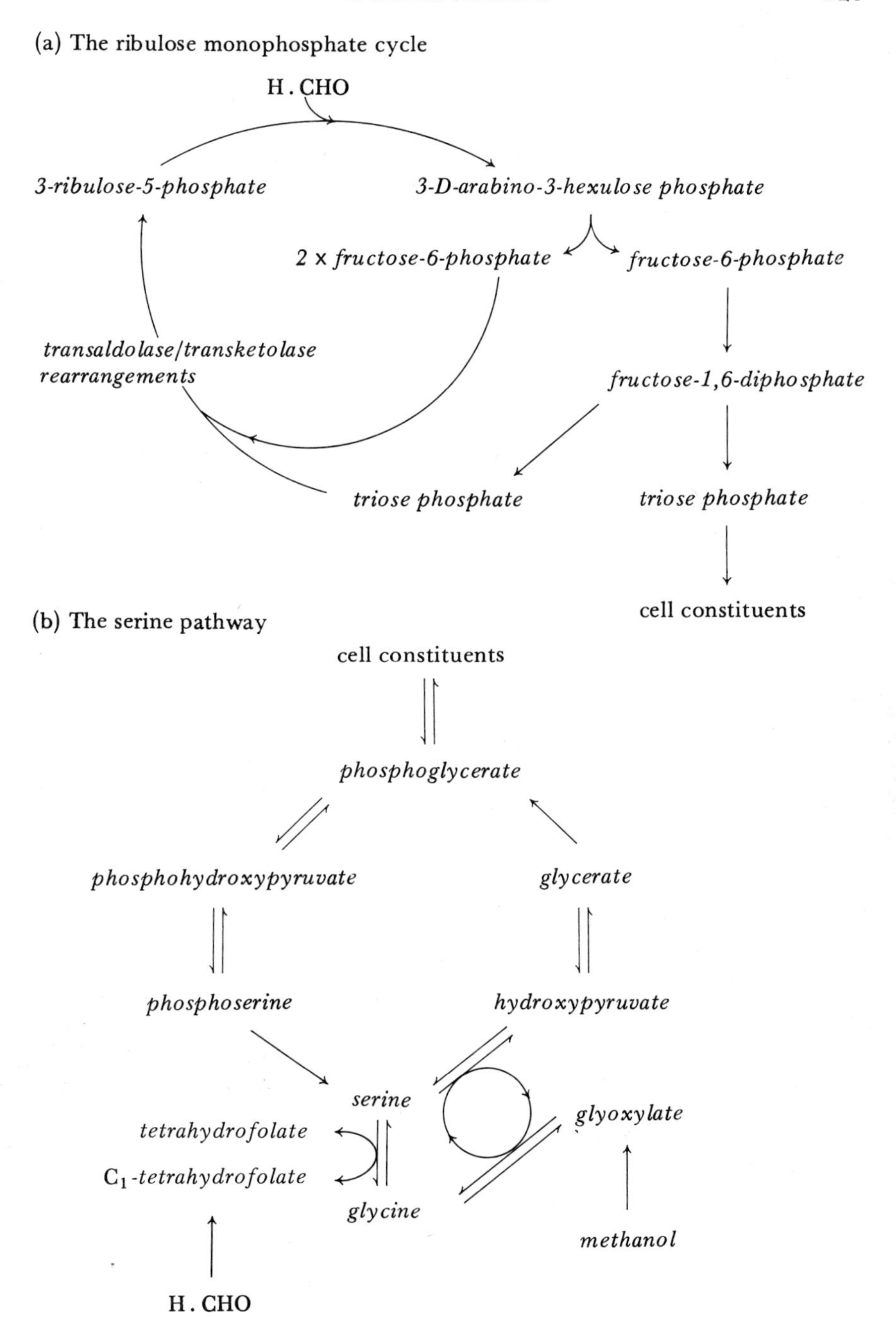

Fig. 4.15. Carbon incorporation pathways in reduced carbon-one utilizing bacteria.

Route A involves the oxidation of one of the terminal methyl groups to a carboxylic acid and requires the activity of a monoxygenase to form a primary alcohol. This is followed by two dehydrogenation steps to form successively, an aldehyde and the fatty acid which is analogous to the methane oxidation shown in Fig. 4.14. The second route (B) involves oxidation of both ends of the molecule to form an α,ω-dicarboxylic acid. Finally subterminal oxidation sometimes occurs forming a ketone (route C). Route A is the most common and enzymes catalysing the initial terminal oxidation have been studied in some detail. They are complex monooxygenases involving three protein components and two different types have been found; one from *Pseudomonas oleovorans* containing rubredoxin (a non-haem iron protein) and the other from a *Corynebacterium* species which involves cytochrome P_{450} (Fig. 4.17).

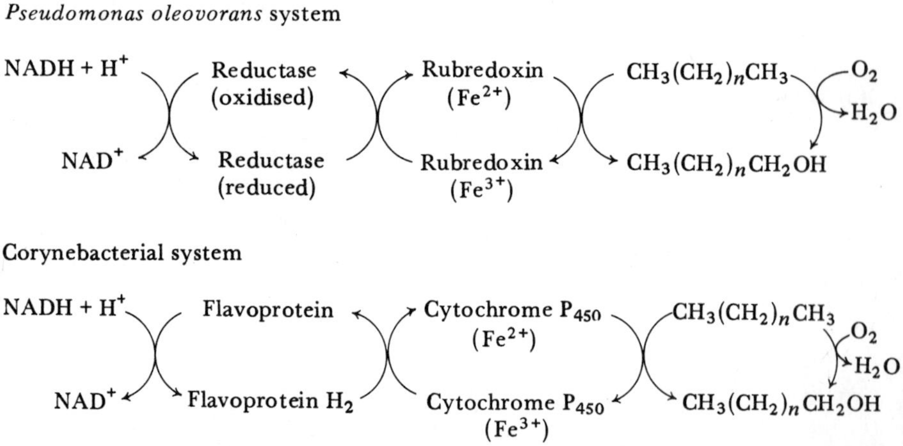

Pseudomonas oleovorans system

NADH + H^+ → NAD^+; Reductase (oxidised) → Reductase (reduced); Rubredoxin (Fe^{2+}) ⇄ Rubredoxin (Fe^{3+}); $CH_3(CH_2)_nCH_3$ + O_2 → $CH_3(CH_2)_nCH_2OH$ + H_2O

Corynebacterial system

NADH + H^+ → NAD^+; Flavoprotein → Flavoprotein H_2; Cytochrome P_{450} (Fe^{2+}) ⇄ Cytochrome P_{450} (Fe^{3+}); $CH_3(CH_2)_nCH_3$ + O_2 → $CH_3(CH_2)_nCH_2OH$ + H_2O

Fig. 4.17. *n*-Alkane terminal oxygenation by microbial monooxygenases.

The alcohol is converted to the corresponding fatty acid by the action of two NAD(NADP) linked dehydrogenases (Fig. 4.18).

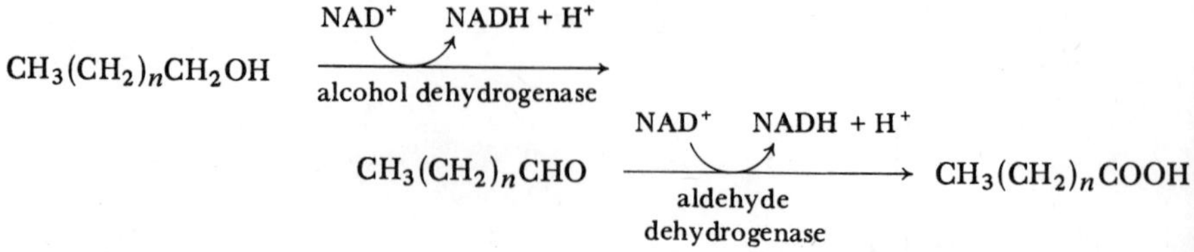

$$CH_3(CH_2)_nCH_2OH \xrightarrow[\text{alcohol dehydrogenase}]{NAD^+ \;\to\; NADH + H^+} CH_3(CH_2)_nCHO \xrightarrow[\text{aldehyde dehydrogenase}]{NAD^+ \;\to\; NADH + H^+} CH_3(CH_2)_nCOOH$$

Fig. 4.18. Microbial conversion of primary alcohols to fatty acids.

The fatty acid is further degraded by conventional β-oxidation to acetate. In the case of branched alkanes, β-oxidation is inhibited at the branch points and this is probably largely reponsible for the slower rates of metabolism of these compounds. There is some evidence that α-oxidation is involved at branch points.

2.3. *Alkenes*

There have been fewer studies of alkene than of alkane degradation. However, alkenes with double bonds in the middle of the carbon chain are probably metabolized in an analogous way to the alkanes. Where the double bond is in the 1-2 position there is evidence, from the yeast *Candida lipolytica*, that an epoxide is formed by a monooxygenase and that this compound is subsequently oxidized to a diol (Fig. 4.19). In some cases it appears that both oxygen atoms in

$$CH_3(CH_2)_nCH{=}CH_2 \xrightarrow[O_2 \quad H_2O]{NADH + H^+ \quad NAD^+} CH_3(CH_2)_n{-}\overset{O}{\widehat{CH{-}CH_2}} \xrightarrow{H_2O} CH_3(CH_2)_n\underset{OH}{CH}CH_2OH$$

Fig. 4.19. Yeast 1-2-alkene oxidation.

the diol come from molecular oxygen. If this is true an epoxide would not be an intermediate but a cyclic peroxide might well be involved. There are also reports of water adding across the double bond and in yet other cases the saturated end of the molecule is attacked first. The various ways in which alkenes are metabolized are summarized in Fig. 4.20.

2.4. *Aromatics*

Microbes have been known to degrade toluene and the xylenes since 1908. Notwithstanding, information about the early steps in aromatic hydrocarbon metabolism is incomplete although much is known about these reactions in the degradation of substituted aromatic compounds such as the phenols, cresols and aromatic acids. Benzene, toluene, naphthalene and methylstyrene are the only aromatic hydrocarbons for which there is much detail. In each case an enzyme catalyses the addition of a molecule of oxygen to the

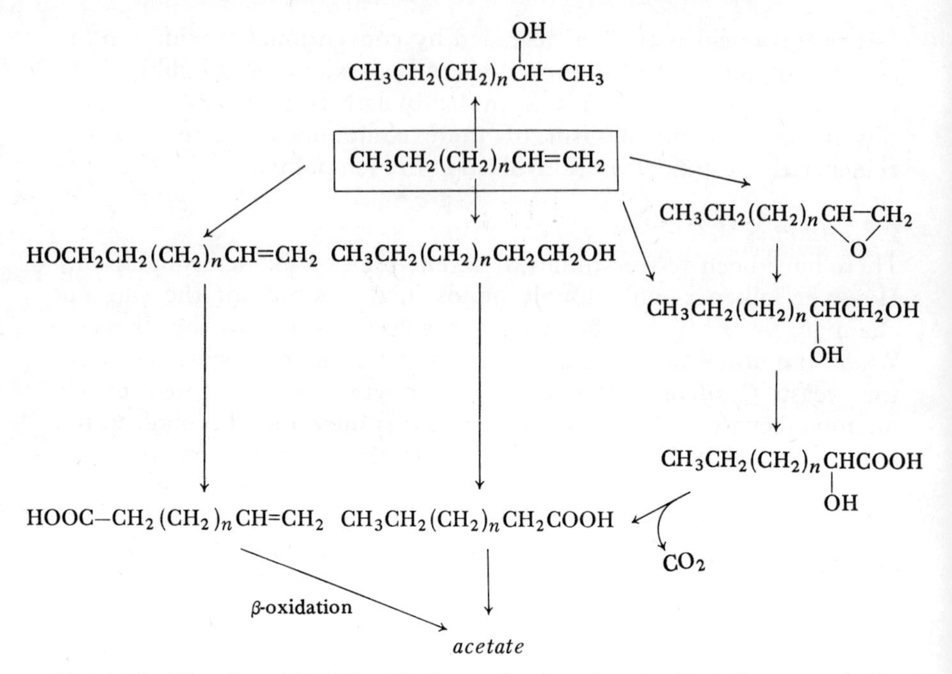

Fig. 4.20. Diversity of initial oxidative mechanisms in microbial alkene metabolism.

aromatic ring to form a diol; in the case of benzene, dihydroxycyclohexadiene is formed which is then converted to catechol (Fig. 4.21).

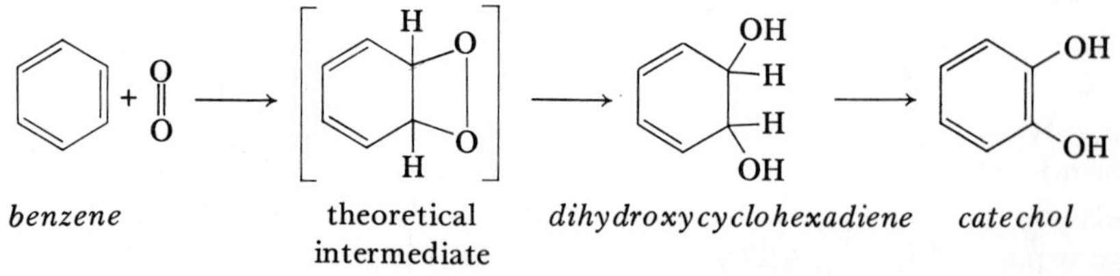

H O O H OH H H OH OH OH

benzene theoretical intermediate *dihydroxycyclohexadiene* *catechol*

Fig. 4.21. Microbial oxidation of benzene.

Catechol is further degraded by one of two routes, depending upon whether the ring is cleaved between the hydroxyl groups ("ortho" cleavage) or adjacent to them ("meta" cleavage). Both reactions are catalysed by dioxygenase enzymes, the former by pyrocatechase (catechol-1, 2-oxygenase), the latter by metapyrocatechase (catechol-2,3-oxygenase). The products are respectively, *cis,cis*-muconate and 2-hydroxymuconic semialdehyde, which are further degraded to normal intermediary metabolites (Fig. 4.22).

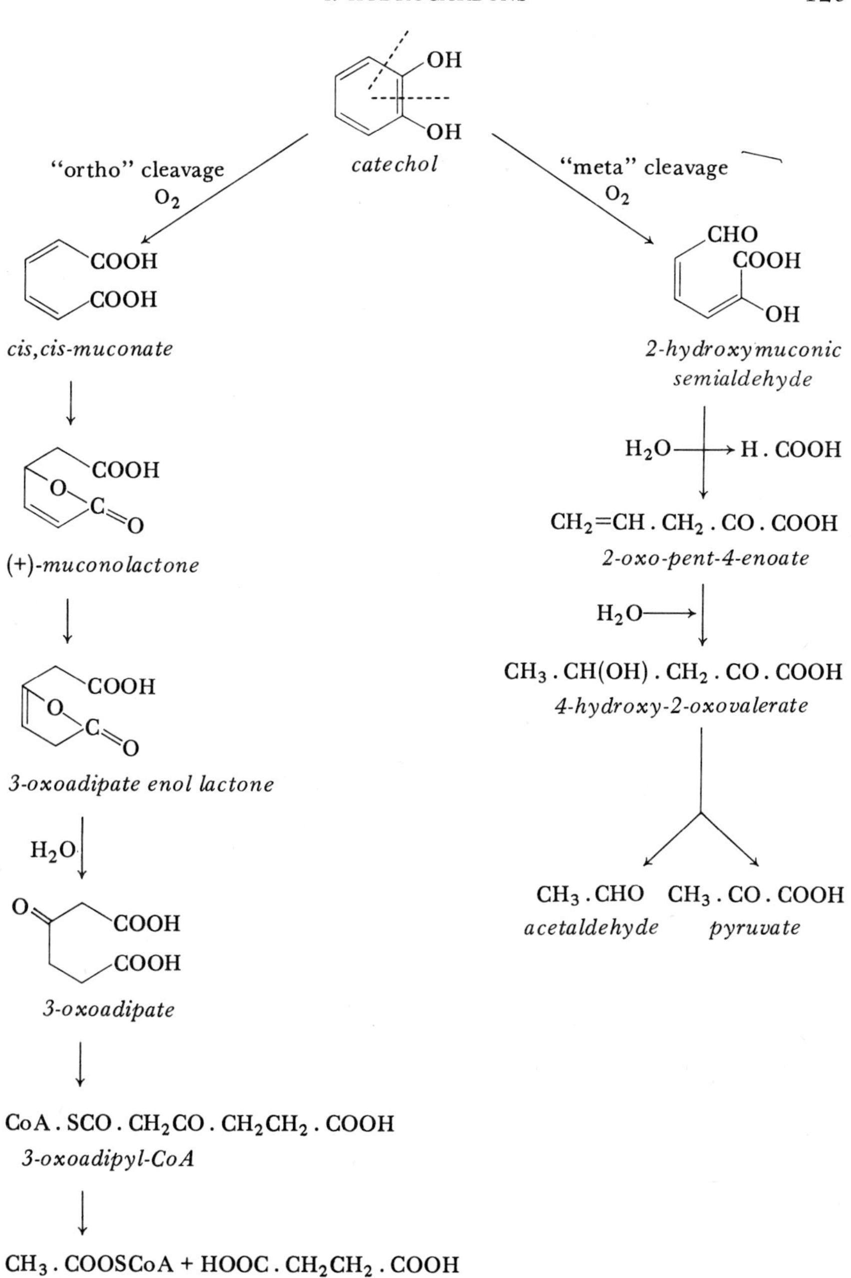

Fig. 4.22. Further metabolism of catechol ring cleavage products.

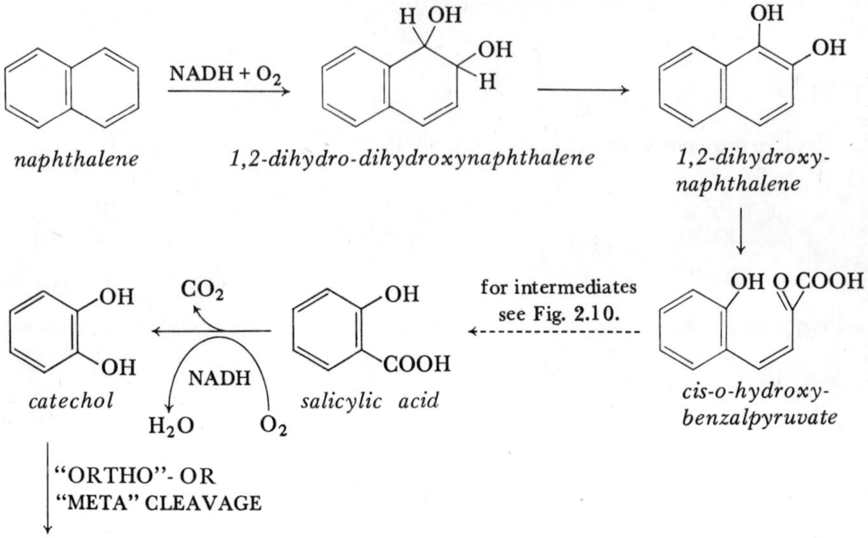

naphthalene — NADH + O_2 → 1,2-dihydro-dihydroxynaphthalene → 1,2-dihydroxy-naphthalene → cis-o-hydroxy-benzalpyruvate ← - - - (for intermediates see Fig. 2.10.) salicylic acid → (NADH, O_2 → H_2O, CO_2) catechol → "ORTHO"- OR "META" CLEAVAGE

Fig. 4.23. Microbial metabolism of naphthalene.

Toluene is metabolized in an analogous way to benzene. The degradation of naphthalene by the sequential metabolism of the constituent aromatic rings is shown in Fig. 4.23. It is thought that the more complex polynuclear aromatic hydrocarbons, such as 3,4-benzpyrene, are degraded in a similar way.

2.5. *Alicyclics*

Alicyclic hydrocarbons are the most resistant of all hydrocarbons to microbial attack. For example, there is to date only one convincing

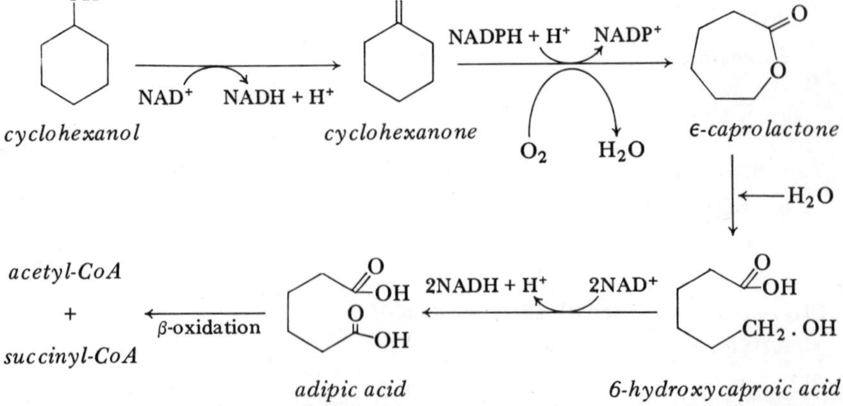

cyclohexanol → (NAD^+ → NADH + H^+) cyclohexanone → (NADPH + H^+ → $NADP^+$; O_2 → H_2O) ε-caprolactone → (H_2O) 6-hydroxycaproic acid ($CH_2.OH$) → (2NAD+ → 2NADH + H^+) adipic acid → β-oxidation → acetyl-CoA + succinyl-CoA

Fig. 4.24. Degradation of cyclohexanol by *Nocardia globerula.*

report of a microorganism that can grow on cyclohexane as its sole source of carbon and energy. Cyclohexane can also be slowly degraded via co-metabolism in a mixed culture of two *Pseudomonas* species. The first of these microbes is an *n*-alkane utilizer which grows readily on heptane. If cyclohexane is also present it is co-oxidized to cyclohexanol which is then degraded by the other member of the mixed culture. In fact several microbes can grow on cyclohexanol (Fig. 4.24).

Nocardia petroleophila grows on methylcyclohexane and the first two steps in its degradation have been described (Fig. 4.25).

CH_3 CH_3 CH_3

OH O

methylcyclohexane *3-methylcyclohexanol* *3-methylcyclohexanone*

Fig. 4.25. Degradation of methylcyclohexane by ***Nocardia petroleophila.***

Several other alicyclic hydrocarbons are known to be co-metabolized.

2.6. Phenylalkanes

The metabolism of phenylalkanes is discussed in detail in Chapter 5 since they are both components and precursors of the alkylbenzene sulphonate surfactants. The degradation of the two components of the molecule basically follow the routes described above for *n*-alkane and aromatic degradation. However, in phenylalkane degradation there is some evidence that α-oxidation is involved in the removal of the alkane carbon atoms near to the ring. This is necessary because the β-oxidation enzyme complex is inhibited by steric hindrance caused by the proximity of the aromatic ring.

2.7. Co-Metabolism

Co-metabolism probably plays an important role in hydrocarbon decay in the environment especially in the case of the more recalcitrant compounds (alicyclics). Co-oxidation of cyclohexane to cyclohexanol is described above and soil microbial isolates have been shown to co-oxidize the series of cycloparaffins from C_3 to C_8 to the corresponding cyclomonoketones. Co-oxidation of higher alkanes by methane-utilizers is also discussed above. In addition, some *Nocardia* strains co-oxidize phenylalkanes to phenylalkanoic acids whilst growing on *n*-alkanes (Fig. 4.26).

Growth substrate: dodecane

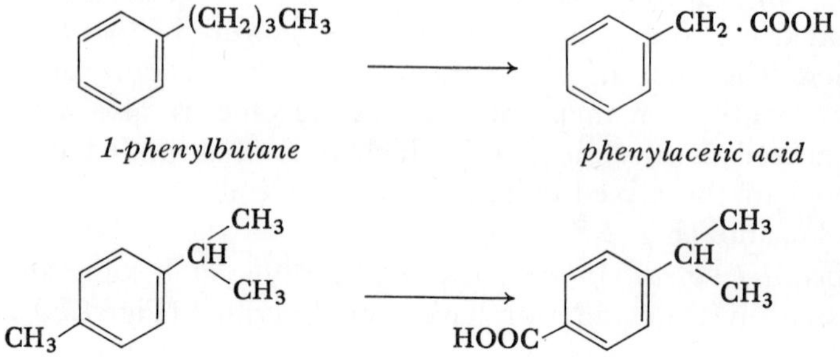

$C_6H_5(CH_2)_3CH_3$ → $C_6H_5CH_2 . COOH$

1-phenylbutane *phenylacetic acid*

$CH_3C_6H_4CH(CH_3)_2$ → $HOOC C_6H_4CH(CH_3)_2$

p-cymene *p-cuminic acid*

Fig. 4.26. Co-oxidation of non-growth hydrocarbons by a *Nocardia* soil isolate.

Simple reactions of this type may be important in converting otherwise recalcitrant hydrocarbons into compounds that can subsequently be metabolized by other microorganisms.

BEHAVIOUR IN THE ENVIRONMENT

Since hydrocarbons include such an array of compounds of widely different physical and chemical properties, the various fractions behave very differently in the environment. Large quantities may be introduced into the biosphere in a number of ways:

A. Local discharge of crude or refined oil fractions onto land or into water. To date any large-scale spills have been due to accidents involving tankers (Torrey Canyon) or to loss of control over off-shore drilling operations (Santa Barbara Channel).

B. Continual widespread emission of crude and refined fractions due to industrial activity and transport.

C. Natural oil seepage from deposits near the surface either on land or at the sea bed.

D. Hydrocarbon release from both living and dead organisms.

The individual contribution that these processes make to the total volume of hydrocarbons in the biosphere is difficult to determine and clearly the quantities involved are not, in themselves the most important consideration. For example it has been calculated that if the cargo of the Torrey Canyon (released when the tanker ran aground and broke up off the Cornish Coast in 1967) had been evenly distributed in the sea at a concentration of 1 ppm it would only have occupied an area of 20 square miles by 500 ft deep. However, this was not what happened and considerable damage was caused by concentrated slicks of oil being carried by the winds and tides. These same natural forces may of course have a beneficial effect and lead to dispersal—particularly where a spill is remote from land. On a world scale, the amount of oil discharged by these occasional accidents is relatively small compared with the continual widespread release due to transportation. It has been calculated that the amount of oil lost into the sea during tanker operations is about 1×10^6 tonnes per annum and that the total lost from all forms of transportation may be greater than this by one or even two orders of magnitude. For example, about 0.0001% (3000 tonnes) of all oil handled at ports is lost to the environment during that handling. Interestingly the amount of hydrocarbon released into the sea from plant life alone is thought to be several million tonnes per annum.

Climate is a factor of prime importance when considering the fate of oil slicks. In this context, there is concern about the use of oil tankers in Arctic seas, especially the use of the North West Passage because oil spilt into these cold waters will be very persistent. The lighter fractions will be slow to evaporate and because of their toxicity to microbes and the low ambient temperatures, biodegradation rates will be extremely low. These combined effects on the breakdown rates are illustrated by Fig. 4.27, which compares biodegradation of fresh Sweden crude oil at various temperatures in seawater collected in the winter to that of similar "weathered" oil from which the lighter fractions have evaporated. This clearly shows that both low temperature and fresh oil dramatically retard degradation. The samples used were not inoculated thus any decay represents the activity of indigenous marine microflora. Samples collected in midsummer showed negligible degradation at 5°C

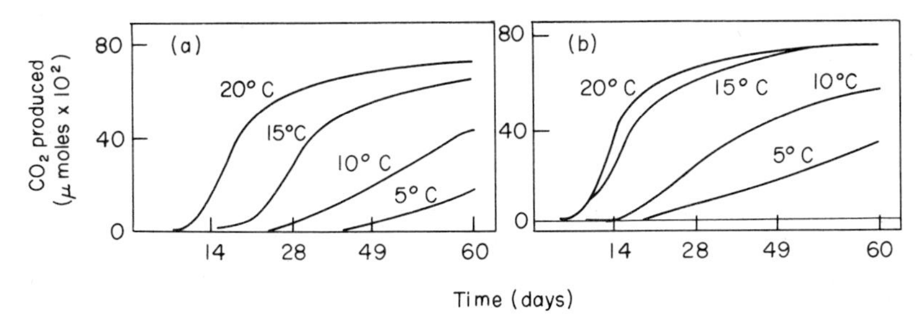

Fig. 4.27. Mineralization of fresh (a) and weathered (b) Sweden crude oil in seawater collected in winter (from Atlas, R. M. and Bartha, R. (1972). *Can. J. Microbiol.* **18**, 1851).

suggesting that the hydrocarbon-metabolizing microflora changes or "adapts" with changes in temperature.

Another environmental factor that greatly reduces the rate of microbial degradation of an oil spill is so-called "chocolate mousse" formation which often occurs when oil is spilt on a rough sea. It is an emulsion of 70-80% seawater in oil.

ECOLOGICAL CONSIDERATIONS

1. Microbial Degradation of Oil Spills

Crude oils spilt on fertile soil in temperate or tropical regions are more or less completely degraded by the microflora in two to six months. However, in poorly developed soils, such as deserts, beaches or in cold climates, where microbial activity is relatively low, the rate of degradation may be measured in years. A significant proportion of low boiling point fractions are lost from the soil by evaporation. Soil molds, yeasts and bacteria have all been shown to be important in oil degradation and are active over a wide temperature range. The quantity of spilt oil responsible for marine pollution is currently estimated to be about 5×10^6 tonnes per year. Evaporation of low boiling point fractions and microbial degradation are the two most important factors in eliminating oil, certainly from the warmer waters. In colder climates, where evaporation is minimal, the lighter fractions which are toxic to microorganisms, may further reduce the already slow rate of biodegradation. Marine bacteria and filamentous fungi are the most important microbial groups involved in oil degradation in the sea. In temperate waters the rate at which these microbes oxidize any large oil spill is almost certainly limited by the local concentrations of essential minerals especially those of nitrate and phosphate.

There have been several investigations of the microbial degradation of crude oil in the laboratory using both pure and mixed cultures. Since crude oil is such a complex mixture it is not an unexpected finding that mixed cultures are more effective at degrading it. The saturated linear alkanes are always utilized preferentially, although those above C_{25} are degraded rather slowly as are the branched-chain isoprenoid hydrocarbons such as pristane [I] and the cyclic components.

$$CH_3CH(CH_3)(CH_2)_3CH(CH_3)(CH_2)_3CH(CH_3)(CH_2)_3CH(CH_3){-}CH_3 \qquad [I]$$

It has not proved possible in the laboratory to demonstrate total degradation of crude oil as there is always a variable amount of recalcitrant material.

2. Biodeterioration of Hydrocarbons

Hydrocarbons are usually susceptible to biodeterioration where there is an oil/water interface. The most important industrial and commercial problems resulting from this are due to the associated microbial growth. For example, the lubricative properties of metal cutting oils are severely reduced by heavy microbial contamination. In addition, slimes in diesel engines and microbial growth in storage tanks and aircraft fuel tanks can lead to blockage of filters and mechanical failure. A variety of biocides have been used to minimize these problems.

3. Effect of Hydrocarbons on Microorganisms

The low molecular weight volatile hydrocarbons, such as hexane and heptane, are quite toxic to a wide range of microbes including some hydrocarbon-utilizers. This is probably because they are very effective lipid solvents and so may destroy or disorganize cell-membrane and wall lipids. Some of the non-hydrocarbon components of crude oil (phenols and substituted alicyclics) also have an inhibitory effect on microbes.

Hydrocarbons in the environment can have a pronounced effect on the local microflora. For example, soils with high oil contents support bacterial populations with high proportions of hydrocarbon utilizers, and indeed this has been used as a means of exploration for underground oil and gas deposits especially in the U.S.S.R. and the U.S.A. The gaseous fraction associated with petroleum is a mixture of methane, ethane and propane. In areas rich in petroleum these gases may seep to the surface and provide substrates for hydrocarbon-utilizing bacteria. Therefore, when one finds high concentrations of these bacteria it is possible that there is a petroleum deposit close by. Looking for methane-utilizers is not practical as methane is produced by many fermentation systems unrelated to petroleum. Ethane, however, is often associated only with petroleum and is a better guide. Some hydrocarbon utilizers fix elemental nitrogen and there have been reports of increased soil fertility in the region of natural gas leaks. The marine microflora can also be enriched by the presence of hydrocarbons. For example, sediments in the Bantaria Bay area of the Gulf of Mexico (where there is chronic oil pollution) have been shown to have up to 1×10^6 oil-degrading bacteria per gram.

4. Effect of Hydrocarbons on Macro-Organisms

4.1. Oil Spills and Birds

The effect of oil spills on birds is the most dramatic and well-documented ecological repercussion of oil pollution. Diving birds, such as puffins and razorbills, are the most vulnerable since they penetrate the oil layer when searching for fish and their feathers become clogged with oil. In addition, oil destroys the waterproofing properties of the feathers so that the birds die of cold or their susceptibility to pneumonia is increased. Furthermore, they cannot feed because they cannot fly, and in attempts to clean their plumage, they preen themselves and imbibe the toxic oil. The most poisonous components are the low boiling point fractions so that light refined fractions or fresh spills are the most dangerous. Unhatched chicks are also poisoned by these lighter fractions and birds usually stop laying after ingesting oil. It is thought that the Torrey Canyon disaster alone killed 100 000 seabirds of which the guillemots were most seriously affected. Although the effects on birds are serious and there is evidence in a few cases of oil causing a long-term decline in population, it seems that in general terms, numbers recover rapidly and that no species is in danger of extinction as a result of oil pollution alone.

4.2. Oil Spills and Other Marine Flora and Fauna

Oil has dramatic effects on many species of marine flora and fauna. For example seaweed, molluscs (limpets, muscles, winkles) and crustaceans are killed by thick layers of oil. In temperate zones marine environments will return to normal often within a year of a major spill.

Another aspect of the problem that is a cause for economic concern is the possible effects of oil pollution on fisheries. The main fishing areas are on the continental shelves which are also the areas with the greatest concentrations of shipping, the sites of off-shore oil fields and outlets for river water from industrial areas. In other words continental shelves constitute the parts of the oceans most subject to oil as well as all other forms of pollution. The effect on fish is very complex, being a summation of the type of oil involved, the habits of the particular species, the climate, etc. The Baltic and North Sea in particular are already fished to the limit and are heavily contaminated so that any detrimental effect of pollution on the marine life will have immediate economic consequences. Fish eggs, developmental stages and young fish are more sensitive to hydrocarbons than the adults and oil can cause abnormal foetal development. Spills can also reduce the food supply to fish by slowing down plankton growth, largely by reducing the light intensity in the surface water.

In the open sea there is little evidence of serious danger to fish but extensive damage has occurred in estuarine waters. These problems are usually more serious in the U.S.A. than elsewhere because a large proportion of the oil in transit is light fractionated material which is relatively toxic. The Caspian sea has also been very badly polluted with oil and the fisheries seriously affected. Aromatic and phenolic components, if not directly harmful to the fish, may give rise to an unpleasant taste even at very low concentrations and it is now common in many areas to put shellfish in clean seawater for several days before consumption in an attempt to wash out the phenolics. There is, at present, little information concerning the public health hazard of consuming fish from oil polluted waters although the main danger is probably due to the ingestion of carcinogenic hydrocarbons such as 3,4-benzpyrene. These polycyclics tend to become concentrated in the fat of marine organisms.

5. Treatment of Hydrocarbon Contamination of the Sea

Various techniques have been applied in attempts to reduce the effects of large scale oil spills. The most widely used method, especially where there is an immediate threat to beaches, involves spraying with non-ionic surfactant emulsifiers. This has proved successful in dispersing the oil, although the surfactants themselves may cause some damage. For example during the Torrey Canyon incident large amounts of detergents were used, most of them being polyethoxylates which are toxic to crustaceans at less than 1 ppm. This treatment disrupted populations of molluscs, crustacea, seaweed and fish but within two years the local ecology had apparently recovered and there is no record of permanent loss of species from the area.

An alternative procedure is to sink the oil by covering it with sawdust and chalk. This is quite effective in that it eventually leads to dispersal but on the debit side may delay biodegradation and foul fishing nets. The oil may be burnt, but it is impossible to obtain complete combustion and this technique is of little value in rough seas due to "chocolate mousse" formation. In enclosed waters, such as harbours and lakes, it is possible to stimulate biodegradation by adding controlled amounts of phosphates and nitrates to enable the water to support a greater mass of hydrocarbon-utilizing microorganisms, although this method has seldom been used. Recent experiments have involved actually applying hydrocarbon-degraders together with water insoluble, yet utilizable sources of nitrogen (paraffinized urea) and phosphorus (octyl phosphate). This is a promising new development especially for final clean-up operations after the bulk of the oil has been removed by more conventional methods.

In protected waters there are several methods for containing oil, the simplest of which involves physical containment of small spills using booms. Chemical "herders" such as oleic and stearic acids, have a similar effect if sprayed around the slick. An ingenious innovation is to spray the oil with a water-insoluble ferro-fluid rendering it magnetic and containable by large magnets.

Decision-making about the methods to be used to treat a particular oil spill are often characterized by conflicts between biological, recreational and industrial interests. For example, after the Torrey Canyon disaster, the oil slicks were serious threats to many resorts and so there was much local pressure to deal with them as rapidly and effectively as possible. In this situation, any possible long-term biological consequences of the procedures chosen were almost certainly not given sufficient consideration as evidenced by the extensive use of detergents which are so toxic to marine life.

CONCLUSIONS

It is clear that there has been a dramatic increase in the release of hydrocarbons into the biosphere over the last two hundred years. Generally speaking they are rapidly degraded by a wide range of microorganisms and do not seem to constitute a serious global pollution problem. Nevertheless in recent years there have been a number of serious problems due to accidents and there are also many examples of increases in local chronic oil pollution. Additional aspects that require careful consideration are the increase in oil contamination of polar waters where degradation is extremely slow and the impending use of tar sand oils which are relatively recalcitrant. It is hoped that these aspects will be given more careful consideration than has been the case in the past.

Recommended Reading

Atlas, R. M. and Bartha, R. (1973). Petroleum in the marine environment. *Residue reviews* 49, 49.

Clark, R. B. (1971). The biological consequences of oil pollution of the sea. *In* "Water Pollution as a World Problem: the legal, scientific and political aspects," p. 53. Europa Publications.

Dagley, S. (1971). Catabolism of aromatic compounds by microorganisms. *Advances in Microbial Physiology* 6, 1.

Floodgate, G. D. (1972). Biodegradation of hydrocarbons in the sea. *In* "Water Pollution Microbiology" (R. Mitchell, Ed.), p. 153. Wiley Interscience, New York.

Klug, M. J. and Markovetz, A. J. (1971). Utilization of aliphatic hydrocarbons by microorganisms. *Advances in Microbial Physiology* 5, 1.

Quayle, J. R. (1972). The metabolism of one-carbon compounds by micro-organisms. *Advances in Microbial Physiology* 7, 119.

5
Surfactants

INTRODUCTION

Surfactants (or surface active agents) are composed of molecules which have both hydrophilic and hydrophobic properties. They therefore tend to aggregate at air-water and oil-water interfaces, reduce surface tension and facilitate emulsification. As a result, these compounds find extensive use as cleaning agents. Soap surfactants (the alkaline salts of fatty acids) were one of Man's earliest synthetic chemical products and there is evidence that their use dates back to the time of the Ancient Egyptians. In the last twenty-five years, however, they have been largely replaced by detergents containing other surfactants.

The reasons for the introduction and commercial success of these new surfactants are two-fold. Firstly, the cheapness of their production from products of the petroleum industry (mainly tetrapropylene and benzene), and secondly the ineffectiveness of soap when used with acid or hard water. In acidic water non-ionized lipophylic fatty acids are formed whilst hard water produces insoluble calcium and magnesium salts. Commercial detergent preparations contain only 10-30% of the active surfactant; the remainder consisting of polyphosphate salts and other ingredients which increase the cleaning efficiency of the product.

Generally speaking synthetic surfactants do not appear to be serious threats to the environment. At low concentrations they are non-toxic to animals and plants, although their surface active properties make them toxic to microorganisms, particularly in the higher concentrations which may occur in sewage plants. It is clear that synthetic surfactants are rapidly removed from the environment, since there has not been a significant increase in the amounts of these compounds in soils, natural waters, plants or animals since 1950 even though world annual production has increased from 35 000 to greater than 9 000 000 tonnes. However, it should be remembered that application of surfactants in high concentrations to alleviate the effect of oil spills may have adverse ecological effects (see p. 137).

In the early days of surfactant use there was considerable concern about the visual pollution caused by foam formation on rivers and lakes. On one celebrated occasion in 1963 the Ohio river in West Virginia was covered from bank to bank with a two foot blanket of foam and it is apparent that under appropriate conditions a surfactant concentration as low as 0.5 ppm can cause extensive foaming. This problem has been reduced by the introduction of linear alkylbenzene sulphonates (LAS) which are more rapidly biodegraded than their predecessors the tetrapropylene-derived alkylbenzene sulphonates (TBS). Although surfactants themselves present little threat and are readily degraded by microbes, some of the non-surfactant additives in commercial detergent preparations have been implicated in environmental problems. Most notably polyphosphates (added as water softeners and to maintain alkalinity) which accumulate in natural waters and may, in part, be responsible for massive algal blooms and eutrophication.

CHEMISTRY

Surfactants can be conveniently classified into three major groups, anionic, cationic and non-ionic. A fourth class, mixed surfactants are also produced in small quantities for special purposes.

1. Anionic Surfactants

Anionic surfactants are compounds which, in aqueous solution, dissociate to give negatively charged surfactant ions, usually derived from sulphonate, sulphate or carboxyl groups.

1.1. Soaps

Until the middle of the twentieth century, soaps were the most important surfactants and even today hold this position in the underdeveloped world. In the industrialized nations, their use is now generally restricted to those situations where mild surfactants are required. For example, tablet soaps, used for personal washing, now represent only about 15% of the total production of surface active agents in the U.S.A. Soaps are composed of the sodium salts of fatty

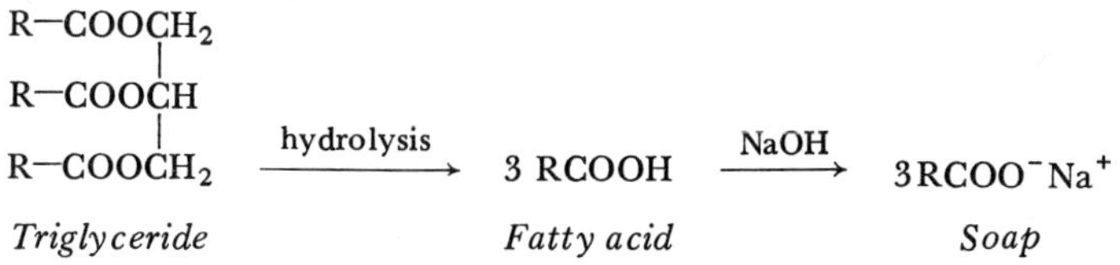

$$
\begin{array}{l}
R{-}COOCH_2 \\
\quad\;\; | \\
R{-}COOCH \\
\quad\;\; | \\
R{-}COOCH_2 \xrightarrow{\text{hydrolysis}} 3\ RCOOH \xrightarrow{\text{NaOH}} 3RCOO^-Na^+ \\
\textit{Triglyceride} \qquad\qquad\quad \textit{Fatty acid} \qquad\qquad \textit{Soap}
\end{array}
$$

Fig. 5.1. Synthesis of soap.

acids and are prepared by neutralizing those fatty acids derived originally from animal or plant oils (triglycerides) (Fig. 5.1). The alkyl group (R) usually contains an odd number of carbon atoms most commonly fifteen and seventeen (fatty acids being biosynthesized from acetate).

1.2. Alkylbenzene Sulphonates

Alkylbenzene sulphonates are the most widely used of the new generation of synthetic surfactants. They are obtained by treating the parent alkylbenzene (prepared from petroleum) with sulphuric acid or sulphur trioxide. This reaction gives rise to the sulphonic acid which is then neutralized, usually with sodium hydroxide, to form the sodium salt (Fig. 5.2).

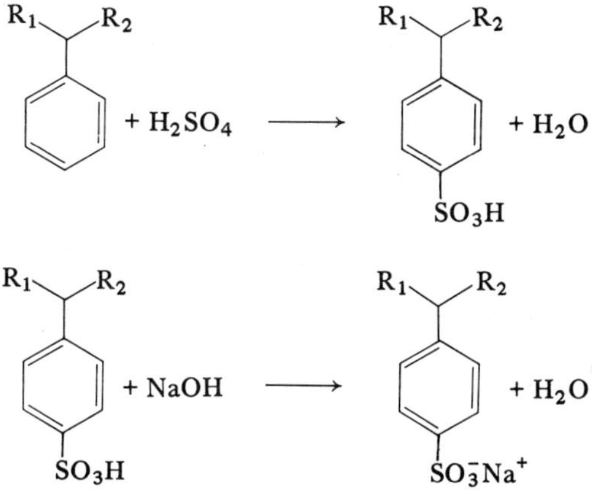

R_1 R_2 + H_2SO_4 ⟶ R_1 R_2 + H_2O SO_3H

R_1 R_2 SO_3H + NaOH ⟶ R_1 R_2 + H_2O $SO_3^-Na^+$

Fig. 5.2. Alkylbenzene sulphonate synthesis.

The alkane chain is usually about twelve carbon atoms in length, R_1 and R_2 are variable but the phenyl group is never terminal in commercial preparations. Most alkylbenzene sulphonates synthesised today have linear alkane moieties, but earlier products from tetrapropylene (TBS) were branched. The sulphonate group enters the ring predominantly in the "para" position.

1.3. Alkane Sulphonates

Alkane sulphonates are more expensive and generally less effective when compared with the alkylbenzene sulphonates. They are prepared by reacting the appropriate alkane(s) with sulphur dioxide and oxygen, followed by neutralization of the resulting sulphonic acid with the chosen base (Fig. 5.4) followed by neutralization.

$$C_nH_{2n+2} + SO_2 + 1/2O_2 \longrightarrow C_nH_{2n+1}SO_3H \xrightarrow{NaOH} C_nH_{2n+1}SO_3Na^+$$

Fig. 5.3. Alkane sulphonate synthesis.

The sulphonate group can add at any position along the chain so that the final product may prove to be a mixture of isomers.

1.4. Olefin Sulphonates

Olefin sulphonates came into commercial use in 1969. Their synthesis involves the reaction of sulphur trioxide with linear α-olefins followed by neutralization (Fig. 5.4).

$$CH_3(CH_2)_n.CH{=}CH_2 + SO_3 \rightarrow CH_3(CH_2)_n.CH{=}CH.SO_3H$$

Fig. 5.4. Olefin sulphonate synthesis.

1.5. Primary Alkyl Sulphates

Primary alkyl sulphates are widely used although they are slightly more expensive to manufacture than the alkylbenzene sulphonates and, for most purposes, no more effective. They are produced by reacting the primary alcohols with sulphuric acid followed by neutralization with the appropriate base (Fig. 5.5).

$$RCH_2OH + H_2SO_4 \rightarrow RCH_2OSO_3H + H_2O$$

$$RCH_2OSO_3H + NaOH \rightarrow RCH_2OSO_3Na + H_2O$$

Fig. 5.5. Primary alkyl sulphate synthesis.

1.6. Secondary Alkyl Sulphates

Secondary alkyl sulphates are prepared from alkenes by treating them with sulphuric acid. For example, when using a linear alkene, the sulphate ester group adds at any position except the terminal carbon atoms so that a complex mixture of isomers results (Fig. 5.6).

$$C{-}C{-}C{-}C{-}C{-}C{-}C{-}C{-}C{-}C{-}C{=}C \xrightarrow[H_2SO_4]{} \underbrace{C{-}C{-}C{-}C{-}C{-}C{-}C{-}C{-}C{-}C{-}C{-}C}_{OSO_3H}$$

Fig. 5.6. Secondary alkyl sulphate synthesis.

Teepol (a common industrial preparation) is an aqueous solution of secondary sodium alkyl sulphates.

1.7. Ester and Amide Sulphonates

Ester and amide sulphonates are rather expensive but there are some special applications for which they are better suited than other anionic surfactants. They are made by reacting acyl chlorides with either short chain hydroxysulphonic salts or aminosulphonic salts (Fig. 5.7).

$$RCOCl + HOCH_2CH_2SO_3Na \longrightarrow RCO_2CH_2CH_2SO_3Na + HCl$$

$$RCOCl + HN(CH_3)CH_2CH_2SO_3Na \longrightarrow RCON(CH_3)CH_2CH_2SO_3Na + HCl$$

Fig. 5.7. Ester and amide sulphonate synthesis.

1.8. Sulpho-fatty Acids

Sulpho-fatty acids are rarely used because they are expensive to manufacture. They are made by sulphonation of the fatty acid with sulphur trioxide followed by the neutralization of the sulphonate and carboxyl groups. The sulphonic group is always found in the α-position (Fig. 5.8).

$$CH_3(CH_2)_nCH_2COOH \xrightarrow{SO_3} CH_3(CH_2)_nCH(SO_3H)COOH \xrightarrow{NaOH} CH_3(CH_2)_n \cdot CH(SO_3Na)COOH$$

Fig. 5.8. Sulpho-fatty acid synthesis.

1.9. Sulphated Polyoxyethylene Ethers

Sulphated polyoxyethylene ethers are derived from the polyoxyethylene non-ionic surfactants and have the general formula [I] were n is usually 6 or 7 (but is greater in non-ionic analogues).

$$R(OCH_2CH_2)_nOSO_3Na \qquad [I]$$

2. Cationic Surfactants

Cationic surfactants dissociate in water to give positively charged ions. They are not used with anionic surfactants as the two types neutralize each other and form water-insoluble salts. They are usually quaternary ammonium derivatives [II] such as dodecyltrimethyl ammonium bromide [III].

$$RN(CH_3)_3^+Cl^- \qquad [II]$$

$$CH_3.(CH_2)_{11}.\overset{(CH_3)_3}{N^+}\,Br^- \qquad [III]$$

Cationic surfactants are not such effective cleansing agents as the anionics, are expensive and so represent only a minor fraction of the surfactant industry. Their main value lies in their powerful bacteriostatic properties.

3. Nonionic Surfactants

Nonionic surfactants, which account for some 25% of the total surfactants in use today, contain hydrophilic groups which do not ionize in solution.

3.1. Alcohol Ethoxylates

Alcohol ethoxylates are produced by reacting aliphatic alcohols with ethylene oxide (Fig. 5.9).

$$ROH + nC_2H_4O \rightarrow R(OC_2H_4)_nOH$$

Fig. 5.9. Alcohol ethoxylate synthesis.

Early preparations of this type were prepared primarily from branched chain C_{13} alcohols but since these are only slowly biodegraded linear alcohols are more frequently used nowadays.

3.2. Alkylphenol Ethoxylates

Alkylphenol ethoxylates are prepared by reacting alkylphenols with ethylene oxide (Fig. 5.10).

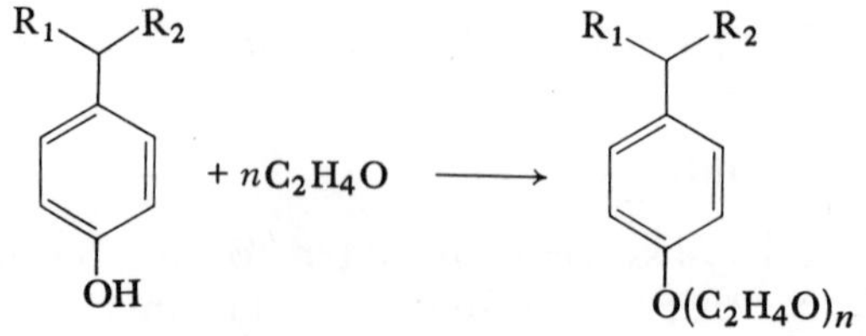

R_1, R_2 (alkyl-substituted phenol, OH) $+ nC_2H_4O \longrightarrow$ R_1, R_2 (alkyl-substituted phenyl) $O(C_2H_4O)_n$

Fig. 5.10. Alkylphenol ethoxylate synthesis.

3.3. Polyoxyethylene Esters

Polyoxyethylene ester surfactants are made by reacting mixtures of naturally occurring fatty acids with ethylene oxide (Fig. 5.11).

$$RCOOH + nC_2H_4O \rightarrow RCOO(C_2H_4O)_n$$

Fig. 5.11. Polyoxyethylene ester synthesis.

3.4. Polyoxyethylene-polyoxypropylene Compounds

Polyoxyethylene-polyoxypropylene compounds [IV] are mixed, high molecular weight polymers with hydrophobic groups derived from both ethylene and propylene oxides.

$$HO(CH_2 . CH_2 . O)_x . (CH_3 . CH_2 . CH_2O)_y . (CH_2 . CH_2 . O)H \qquad [IV]$$

3.5. Miscellaneous

Other nonionic surfactants manufactured in lesser amounts are shown in Fig. 5.12. Sucrose esters have been proposed as possible "third-generation" surfactants to replace the linear alkylbenzene sulphonates.

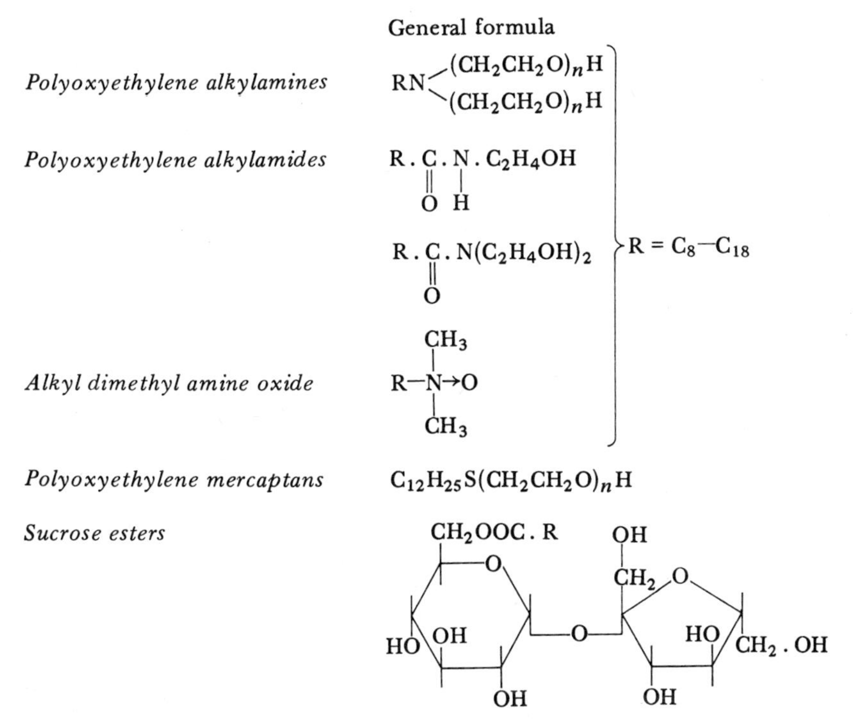

Fig. 5.12. Miscellaneous non-ionic surfactants.

ANALYTICAL METHODS

One of the major problems in examining any habitat for detergents (or indeed any other pollutant) is that of representative sampling. The surface active properties themselves make this difficult because surfactants tend to concentrate at interfaces. Therefore foam contains a much higher concentration of surfactant than the bulk of a solution and should be allowed to subside before sampling. In addition, adsorption occurs on the walls of laboratory glassware causing inaccuracies and cross-contamination of samples. To some extent this problem can be overcome by coating glassware with a silicone water-repellant.

It is often necessary to measure surfactants in complex organic samples (such as sewage sludge) where the biological components may immobilize significant amounts of surfactant. As a result, the apparent concentration of surfactant in the liquid phase may well be much lower than the true concentration. This is especially true with an assay of the alkylbenzene sulphonates when it is necessary to desorb the bound surfactant by extraction with hot water, ethanol or methanol. For accurate results and high recovery rates the sample is first dried, then powdered and finally extracted with an alcohol for twenty hours. Once the analyst is content that his extraction is successful there are a variety of methods available for measuring concentration. These may be classified into four groups: procedures depending on specific surfactant properties, specific chemical methods, physico-chemical methods, and miscellaneous, non-specific methods.

1. Specific Surfactant Properties

1.1. Foaming

Foaming has been a helpful guide is many studies of biodegradation particularly for non-ionic surfactants for which there is no satisfactory chemical method. The foaming phenomenon is extremely complex and is the result of interactions between a host of parameters. Consequently, foam measurements are necessarily very crude and it is quite impossible to use this method to determine exact concentrations of surfactants in natural waters. The amount of foaming, in terms of both volume and persistence, depends upon temperature, aeration, agitation, the shape of the container, etc. In addition, synthetic surfactants may not be the only components of the sample to cause foaming, for many natural substances, particularly proteins, are very effective in this regard. Nevertheless, measuring the amount of foaming under standard conditions can be useful as a semi-quantitative indicator of surfactant concentration. In a typical test, 150 ml of the test liquid is placed in a 7 x 50 cm tube and aerated by 100 ml min^{-1} of air introduced through a sintered glass diffuser at the base. The foam height is measured after five minutes.

1.2. Surface Tension

Surfactant molecules tend to congregate at the surface of aqueous samples and in so doing lower the surface tension. Measurement of these surface tension changes can give a qualitative guide. Surface tension is usually determined using a du Nouy tensiometer which assesses the force required to lift a circular wire ring, 2 cm in diameter, from beneath the liquid through the surface layer. It is most important to allow time for equilibration, particularly after decanting the top of the liquid, otherwise the result will be an underestimate of the amount of surfactant present.

2. Specific Chemical Methods

2.1. Methylene Blue Test

The Methylene Blue Test is broadly applicable to anionic surfactants and is extremely sensitive (0.1 ppm of alkylbenzene sulphonates).

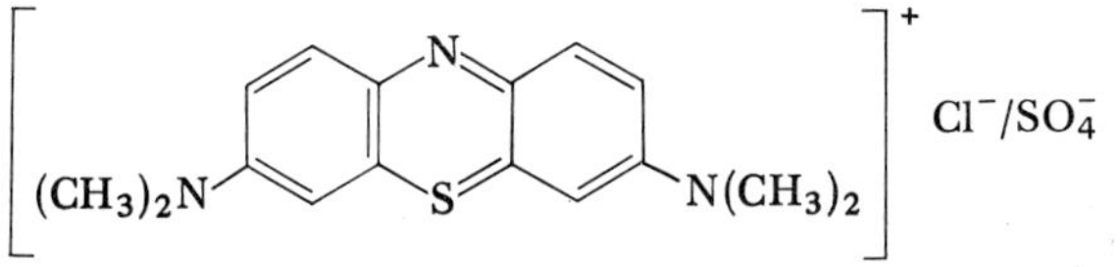

$$\left[(CH_3)_2N\text{–(phenothiazinium: N, S)–}N(CH_3)_2\right]^+ \quad Cl^-/SO_4^- \qquad [V]$$

Methylene blue is a cationic dye [V] which is insoluble in organic solvents. However, it forms salts with anionic detergents which are then readily soluble both in benzene and chloroform. As a result, the measurement of blue colour in the solvent is indicative of the amount of anionic surfactant present in the original sample. The method has to be applied with some caution, for high concentrations of chloride ions tend to displace the methylene blue into the organic layer (a 1.79% NaCl solution gives the same methylene blue response as 10 ppm of alkyl sulphate) and so care is especially necessary with sea water samples. In contrast, some substances actually reduce colour formation, particularly amines and proteins, which successfully compete with methylene blue for salt formation with the surfactant. Fortunately these problems can be resolved by appropriate clean-up procedures. Correction may also need to be applied for the variable responses induced by pH.

A surfactant's response to methylene blue is chiefly an indication of the number of molecules present, so that in order to express the results in terms of weight it is necessary to have a standard such as Manoxol OT (sodium bis (2-ethylhexyl) sulpho-succinate) or sodium myristyl sulphate.

2.2. Estimation of Free Sulphate

The release of sulphate from sulphate- or sulphonate-containing surfactants can be used to indicate the destruction of surfactant properties although it does not indicate whether or not the organic part of the molecule has been affected. There are two commonly used procedures for measuring the amount of sulphate, the first involving turbidimetric determination of barium sulphate; the second the assay of ^{35}S released from labelled surfactant. The former is suitable for surfactant concentration from 1 to 10 ppm. Barium chloride is added to the sample under standard conditions and the resulting turbidity due to barium sulphate formation measured in a spectrophotometer. The radioactive assay is more sensitive and usually involves counting precipitated ^{35}S sulphate.

2.3. Estimation of Non-ionic Surfactants

Chemical methods employed to measure the concentration of the polyethoxylate surfactants usually involve complex formation with heavy metals (such as cobalt, molybdenum or mercury). The assay is complicated by the chemical range of polyethoxylates and the need for elaborate purification of samples to avoid interference from naturally occurring compounds. However, what is known as the cobalt-thiocyanate method, has been widely studied and used to a limited extent in the field. A complex is formed with the polyethoxylate hydrophilic group which is subsequently extracted with an organic solvent and the amount estimated from the absorption either in the visible region at 620 nm or in the ultra violet at around 320 nm.

3. Physicochemical Methods

3.1. Spectroscopy

Infra-red spectroscopy has been used to follow the biodegradation of both alkylbenzene sulphonates and non-ionic surfactants. The strong absorption bands of the benzene ring in the far ultra-violet (193 and 223 nm) are also valuable for measuring the later stages of alkylbenzene sulphonate degradation when the aromatic ring is being metabolized.

3.2. Chromatography

Both thin layer and paper chromatography have been particularly useful for separating non-ionic surfactants and their biodegradation products, whilst gas liquid chromatography has been extensively used for similar work with the anionic detergents. Samples for GLC analysis need to be pretreated in order to desulphonate the anionics as the surfactant molecules themselves are insufficiently volatile. This involves heating with concentrated phosphoric acid at 2000°C to release the corresponding alkylbenzene. GLC is also useful for the detection of carboxylated intermediates formed during biodegradation and may be used after the samples have been treated with a methylating agent (e.g. diazomethane) to form volatile methyl esters.

4. Miscellaneous Methods

4.1. Radiotracer Techniques

Radio-labelled detergents have been used extensively in biodegradation studies. ^{35}S alkylbenzene sulphonates, as mentioned previously, are very useful for studies in the field making it possible to detect extremely small amounts of these surfactants or their breakdown products. Carbon-14 and tritium labelled surfactants have also been used.

4.2. Chemical Oxygen Demand

Chemical oxygen demand (COD) measures the amount of oxygen required to oxidize the surfactant (or other organic substances, Chapter 3) in a sample. The determination involves boiling with excess standard dichromate and back-titrating with ferrous ammonium sulphate. The method is insufficiently sensitive for surfactant concentrations below 5 ppm.

4.3. Combustion

Combustion also measures the total organic material. The sample is burnt in a stream of oxygen and converted to carbon dioxide which is then measured by infra-red spectroscopy. This does not, of course, give exactly the same information as the chemical oxygen demand since it indicates the total amount of carbon in the sample and not its oxidation state.

4.4. Biochemical Oxygen Demand

Biochemical oxygen demand (BOD) is a measure of the microbial oxygen consumption resulting from the metabolism of the surfactant. The "classical" method (Chapter 3) involves filling a bottle with an aqueous solution containing the test compound along with a known amount of oxygen and microbial inoculum. The bottle is then left for a standard time (usually five days) before chemically measuring the amount of oxygen left. It is also possible to monitor the oxygen consumption by using an oxygen electrode.

4.5. Other Gas Measurements

Oxygen consumption by microbes, in the presence of test material, can be monitored directly using a respirometer. It is also possible to measure the amount of carbon dioxide respired.

4.6. Microbial Growth

The growth of microbes at the expense of the surfactant can be used as a measure of biodegradation although its occurrence does not necessarily indicate total mineralization to carbon dioxide and water.

MICROBIAL DECAY

Microorganisms are almost entirely responsible for the disappearance of surfactants from the environment. As the main biological components of activated sludge (Chapter 3) they play a critical role in the degradation of detergents in sewage treatment plants. Two major types of investigation have been used to elucidate the metabolic routes involved in the degradation of surfactants. Firstly the use of so-called "die away tests", in the simplest of which the surfactant is dissolved in a sample of river water and the solution examined at intervals. Fig. 5.13 shows typical patterns obtained for LAS and TBS. The second type of investigation, yielding more detailed biochemical information, involves studies of the metabolism of single surfactant species by pure cultures of microorganisms.

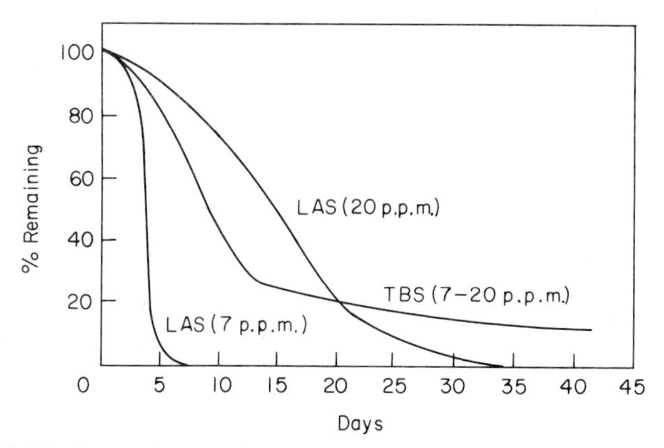

Fig. 5.13. Degradation of alkylbenzene sulphonates in river water.
From: Swisher, R. D. "Surfactant-Biodegradation" 1970, Marcel Dekker, New York.

1. Anionic Surfactant Degradation

There is more information available concerning the metabolic routes involved in the degradation of the anionic surfactants than for the other two groups, because of their commercial value and the availability of suitable methodology.

1.1. Soaps

Soaps are not significantly different in their biodegradative properties when compared with modern commercial surfactants. To a large extent their environmental acceptability has been due to their ready precipitation as insoluble calcium and magnesium salts, and not to any special susceptibility to biodegradation. Nevertheless, soaps are metabolized quite rapidly by microorganisms, chiefly through β-oxidation. Both soaps and linear alkane sulphonates are degraded in activated sludge at rates of 1-2 mg g^{-1} . $hour^{-1}$. The rate of soap degradation decreases with increasing chain length from C_8 to C_{20} and usually the unsaturated ones are degraded more readily than those that are saturated. In general, even those microorganisms which do not break down soaps can tolerate higher concentrations (>1000 ppm) than they can of other surfactants.

1.2. Alkane Sulphonates

The commercial alkane sulphonates are mainly linear and secondary, with the sulphonate group attached randomly along the chain. These compounds are readily biodegraded but their common impurities, principally non-linear analogues, are metabolized more slowly. "Die-away" tests show that pure compounds of chain length C_{13} to C_{19}

take four to eleven days to biodegrade in river water. There have been few detailed biochemical studies of the degradation of these compounds but it is clear that *Pseudomonas* isolates, growing on short chain *n*-alkane-1-sulphonates, dissimilate these compounds by an initial desulphonation followed by β-oxidation. *n*-Pentane-1-sulphonate is metabolized as shown in Fig. 5.14.

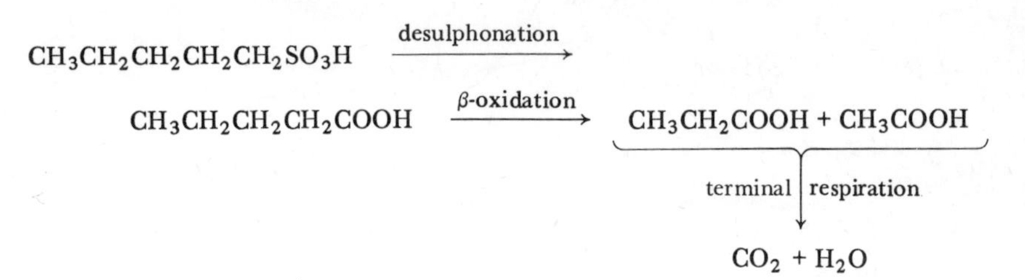

Fig. 5.14. Biodegradation of *n*-pentane-1-sulphonate.

1.3. Olefin Sulphonates

α-Olefin sulphonate surfactants usually contain approximately equal amounts of alkene sulphonates and hydroxyalkane sulphonates. The former component is degraded more rapidly than the latter. There is little information concerning the routes of biodegradation of these compounds, but they are probably analogous to those of the alkane sulphonates.

1.4. Primary Alkyl Sulphates

The linear primary alkyl sulphates are speedily degraded by bacteria. The initial attack, to yield inorganic sulphate and the appropriate alcohol, is catalysed by sulphatase enzymes induced by the substrate. The alcohol is then oxidized, first to the corresponding aldehyde and then to the carboxylic acid by the action of two constitutive dehydrogenases (Fig. 5.15). This pathway is similar to the microbial degradation of alkane hydrocarbons (Chapter 4). The fatty acid is further metabolized by β-oxidation.

$$\mathrm{R.CH_2OSO_3Na} \xrightarrow[\downarrow\ \mathrm{Na_2SO_4}]{} \mathrm{RCH_2OH} \longrightarrow \mathrm{RCHO} \longrightarrow \mathrm{RCOOH}$$

Fig. 5.15. Biodegradation of primary alkyl sulphates.

1.5. Secondary and Branched Alkyl Sulphates

Both the branched and the secondary alkyl sulphates are biodegraded less rapidly than are the primary compounds. The further metabolism of the resulting alcohols is probably similar to that experienced by the primary alcohols. The microbial sulphatase enzymes which hydrolyse the primary alkyl sulphates are inactive with the secondary and branched analogues but some microbes are able to produce sulphatases that are specific for these compounds. In *Aerobacter cloacae* linear secondary alkyl sulphates induce sulphatases which are only active against these compounds.

1.6. Ester and Amide Sulphonates

Commercial preparations of ester and amide sulphonates are made from naturally occurring fatty acids and are readily biodegraded. They are hydrolysed to fatty acids by non-specific microbial esterases and amidases and then β-oxidized to acetyl-CoA.

1.7. Alkylbenzene Sulphonates

Since alkylbenzene sulphonates are the most widely used of all synthetic surfactants their biodegradation has attracted extensive investigation. Some of this work has involved studies of the hydrocarbon precursors of these surfactants, namely the alkylbenzenes. These compounds are not surfactants and are water insoluble, but nevertheless their metabolism can be considered a model.

In some microbes the alkylbenzene is probably absorbed and metabolized following extracellular desulphonation of the surfactant. In the case of both the hydrocarbons and the surfactants, the initial point of attack is at the terminal methyl group of the alkyl side chain. This reaction is catalysed by an alkane monoxygenase (Fig. 5.16). X is usually NAD or NADP and the complex enzyme systems

$$RCH_2CH_3 + O_2 + XH_2 \longrightarrow RCH_2CH_2OH + H_2O + X$$

Fig. 5.16. Enzymic alkane oxidation.

(discussed in greater detail in Chapter 4) contain either cytochrome P-450 or rubredoxin. The alcohol formed in this reaction is then converted, via the aldehyde, to the corresponding carboxylic acid by two dehydrogenase enzymes (Fig. 5.17).

$$R.CH_2.CH_2.OH \xrightarrow{NAD^+ \;\to\; NADH + H^+}$$

$$R.CH_2.CHO \xrightarrow[H_2O]{NAD^+ \;\to\; NADH + H^+} R.CH_2COOH$$

Fig. 5.17. Enzymic oxidation of long chain alcohols.

The capacity to oxidize a methyl to a carboxyl group in this way is widespread amongst microorganisms. Since the initial step involves the incorporation of molecular oxygen, the process is dependent on aerobic conditions. The next step in the metabolism of both the alkylbenzene sulphonates and the hydrocarbon analogues involves β-oxidation of the alkanoic acid portion to acetyl-CoA units. Typically, microbes growing on these compounds have elevated levels of the key enzymes of the glyoxylate cycle (isocitrate lyase and malate synthase) because they are growing primarily on C-2 units. They are then faced, like all C-2 utilizers, with the necessity of replenishing their supply of C-4 compounds (which are used for biosynthesis) since acetyl-CoA is oxidized completely to carbon dioxide by the tricarboxylic acid cycle. The glyoxylate cycle serves this replenishing (anaplerotic) function. There is considerable evidence that β-oxidation is not the only process involved in side chain degradation. Odd carbon atom intermediates are often detected from even carbon atom substrates and vice versa suggesting involvement of an α-oxidation mechanism.

Commercial detergent preparations contain all the alkylbenzene sulphonate isomers except the 1-phenyl. In all cases the alkyl chain is probably broken down to yield either benzoic acid or phenylacetic acid (Fig. 5.18). Usually, by this stage, the sulphonate group has been removed.

The rate at which a particular alkyl group is degraded increases with the distance of the benzene ring from the terminal methyl group. This is described as the Distance Principle.

Some pure cultures of bacteria that grow on alkylbenzene sulphonates only degrade the alkyl side chain although most pure and

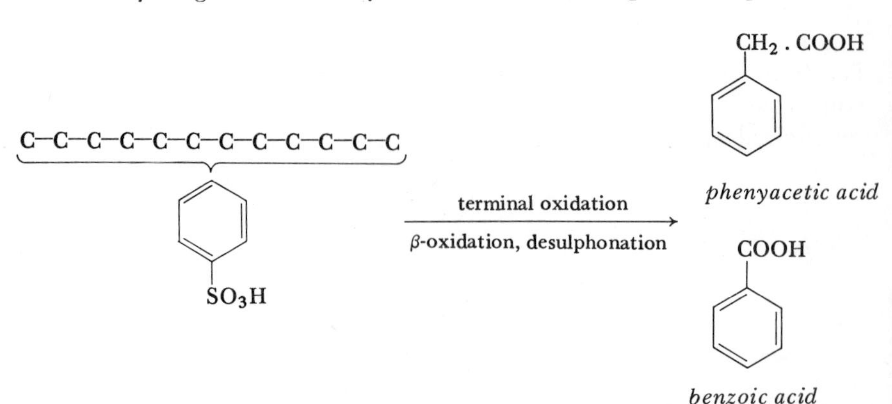

Fig. 5.18. Microbial degradation of alkylbenzene sulphonates.

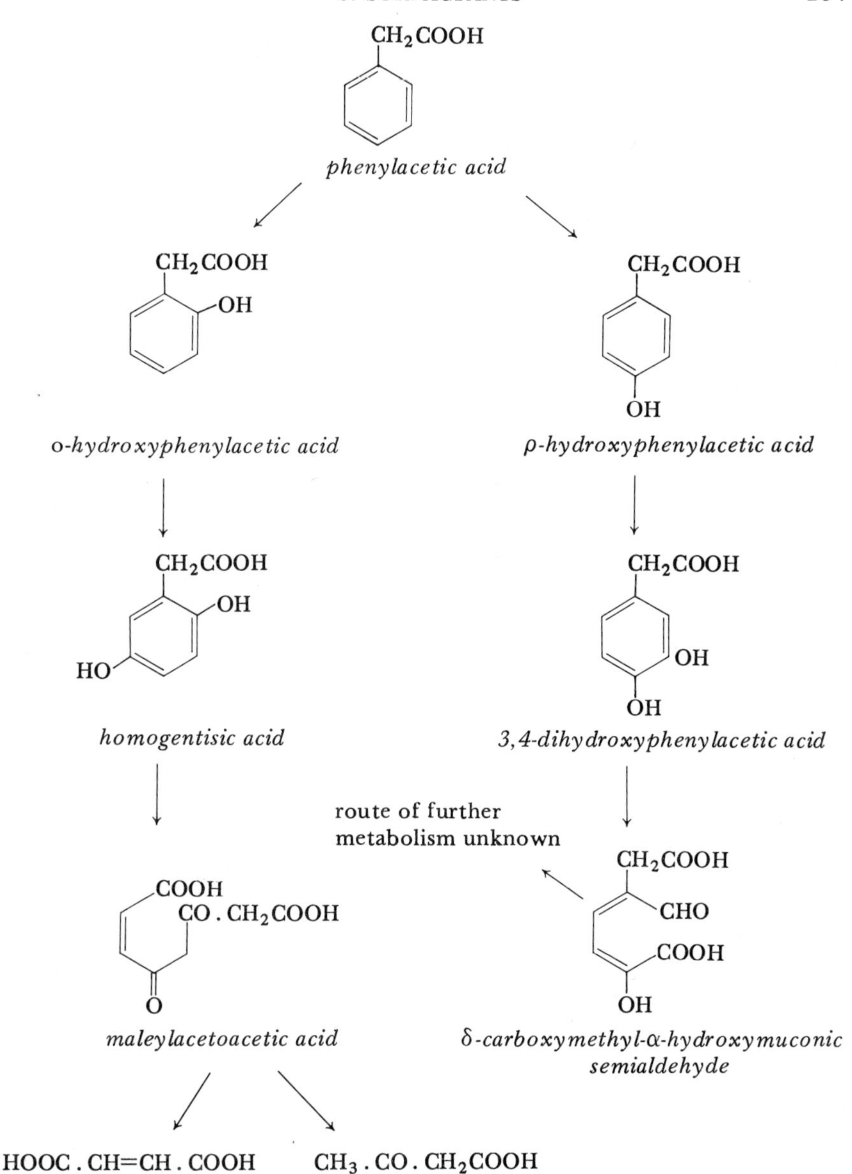

Fig. 5.19. Pathways of phenylacetic acid degradation by bacteria.

mixed cultures also degrade the aromatic ring. The products formed after degradation of the side chain and desulphonation (benzoate and phenylacetate) are further metabolized to diphenols (e.g. catechol) by monooxygenases before the ring is cleaved by a dioxygenase. Depending upon the particular organism or enzyme the cleavage can occur adjacent to one of the hydroxyls ("meta" cleavage) or between them ("ortho" cleavage). The ensuing catabolic routes of the cleavage products to common intermediary metabolites are different. The degradative pathways for benzoate are via catechol and are shown in Chapter 4 Fig. 4.22. In the case of phenylacetate there are separate pathways via the *ortho*- or *para*-hydroxy derivative (Fig. 5.19). Since the enzymes of these pathways are inducible, their activities are typically stimulated in organisms metabolizing alkylbenzene sulphonates.

The removal of the sulphonate group from the alkylbenzene sulphonates is an important step in their biodegradation. This group is far more stable biologically than the sulphate in the alkyl sulphate surfactants previously discussed and there is still uncertainty about the mechanism of its removal. The most detailed study so far has been with a species of *Bacillus* in which two enzymes are involved in the formation of sulphate, the desulphonating enzyme itself and a sulphite-cytochrome *c* oxidoreductase. The immediate product is most probably sulphite which can then be oxidized to sulphate. The mechanism for the desulphonation reaction is not throughly understood, but in some instances it may involve a simple hydroxylative

$$RSO_3H + H_2O \longrightarrow ROH + 2H^+ + SO_3$$

Fig. 5.20. Hydroxylative desulphonation of alkylbenzene sulphonates.

attack on the sulphonate group (Fig. 5.20). It is also possible that the enzyme catalyses a monooxygenation (Fig. 5.21). In some organ-

$$RSO_3H + O_2 + NADH + H^+ \longrightarrow ROH + H_2O + SO_3^- + NAD^+$$

Fig. 5.21. Monooxygenase mechanism of desulphonation.

isms, however, the product is a non-hydroxylated phenylalkanoate indicating that a reductive mechanism is involved (Fig. 5.22).

$$RSO_3H + NADH + H^+ \longrightarrow R{-}H + NAD^+ + H_2SO_3$$

Fig. 5.22. Reductive mechanism of desulphonation.

The stage in biodegradation at which desulphonation occurs probably varies between different microorganisms. There is good evidence that, in several cases, desulphonation occurs extracellularly at the outer surface of the organism and that it is the corresponding phenylalkane that is taken up and metabolized.

Microbes which degrade alkylbenzene sulphonates can be divided into various groups on the basis of the type and extent of metabolism:

A. ω- and β-oxidation of the side chain without either desulphonation or ring metabolism.

B. ω- and β-oxidation of the side chain together with hydroxylative desulphonation, ring cleavage and the further metabolism of ring cleavage products.

C. As in B but with reductive desulphonation forming the phenylalkanoates rather than the *p*-hydroxyphenyl derivatives.

D. As in B but involving α-oxidation as well as β-oxidation and so giving rise to both odd- and even-carbon number intermediates from either odd or even-carbon number substrates.

E. As in D but without ring cleavage.

F. Degradation of branched side chains (occurring in TBS) together with desulphonation and ring cleavage.

G. Initial desulphonation and oxygenation of the aromatic ring forming alkylcatechols and then "meta" cleavage of the ring. This occurs with alkylbenzene sulphonates with short side chains (C_2-C_5).

H. Degradation by co-metabolism, the rate being increased by analogue enrichment with phenol (see Chapter 2 for a definition of co-metabolism).

An exciting aspect of recent research into microbial alkylbenzene sulphonate metabolism is revealed by *Pseudomonas putida* in which it seems that one, and possibly more of the enzymes involved is coded for extrachromasomally. This was first suspected when it was found that pure cultures of microorganisms frequently lost their ability to degrade alkylbenzene sulphonates after prolonged subculture on rich media. This is similar to the loss of drug resistance factors in enteric bacteria. The genes for the desulphonating enzyme and for catechol-2,3-oxygenase (the aromatic ring cleaving enzyme) may be located on a plasmid in this particular *Pseudomonas* species.

As mentioned earlier in this chapter the first commercial alkylbenzene sulphonates, which were derived from tetrapropylene, had extensive branching in the alkyl side chain and posed a considerable problem because they are somewhat resistant to elimination from the environment. Although these compounds are slowly broken down by microorganisms, the rate is much less than for the linear analogues. Additionally, adsorption onto soil or activated sludge removes the surfactant from a degrading environment and is particularly significant in the case of the branched chain compounds. Experiments using ^{35}S-labelled TBS in soil lysimeters have shown that under appropriate conditions 95% primary degradation (minimum degradation necessary to change the identity of the compound) occurs. In activated sludge, *in situ*, TBS is degraded slowly but the biochemistry has not been studied as extensively as has been the case with LAS. Notwithstanding, the information available suggests that the pathways are very similar.

2. Non-ionic Surfactant Degradation

Both the hydrophilic and hydrophobic groups of non-ionic surfactants are organic and hence vulnerable to biodegradation. The most common non-ionic surfactants are ethoxylates of alcohols or alkylphenols and the metabolism of the alcohol moieties (ethylene glycol itself and the polyglycols) has been studied. The results, if interpreted with caution, are pertinent to the metabolism of the corresponding surfactants. The relationship between these compounds and the surfactants is reminiscent of that between the alkylbenzenes and the alkylbenzene sulphonate surfactants. Work with these alcohols has been hampered by a lack of sensitive methods for assay of the parent compounds but nevertheless BOD and respirometric measurements show extensive aerobic and anaerobic biodegradation in most cases. Species of *Pseudomonas, Gluconobacter* and *Acetobacter* act on polyglycols by oxidizing a free terminal hydroxyl group first to an aldehyde and then to the carboxylic acid. Under anaerobic conditions, *Aerobacter aerogenes* converts ethylene glycol to acetaldehyde and propylene glycol to propionaldehyde (Fig. 5.23).

$$HO.CH_2.CH_2OH \longrightarrow CH_3CHO$$

Fig. 5.23. Oxidation of ethylene glycol by *Aerobacter aerogenes*.

The next step probably involves the oxidation of one half of the aldehyde and the reduction of the other (i.e. fermentation occurs). In the case of ethylene glycol, *Clostridium glycolicum* ferments the carbon source, via acetaldehyde, to an equimolar mixture of ethanol and acetic acid.

2.1. Alcohol Ethoxylates

The first stage in the biodegradation of linear primary alcohol ethoxylates involves cleavage of the molecule into the alcohol and polyethylene glycol. Further metabolism probably requires dehydrogenation followed by β-oxidation of the resulting carboxylic acid. The linear secondary analogues are fairly rapidly degraded in a similar manner, whereas the highly branched tetrapropylene-derived compounds are biodegraded only very slowly. In comparative experiments in soil the linear secondary compounds were completely degraded in 14 days, whilst 30% of the highly branched tetrapropylene analogues were unchanged after 49 days.

2.2. *Alkylphenol Ethoxylates*

Complete degradation of alkylphenol ethoxylates does not occur as readily as with their alcohol analogues. The ethoxylate side chain is metabolized by a variety of microbes whilst it is still attached to the hydrophobic part of the molecule. The chain is probably shortened, one ethoxylate group at a time, but the detailed mechanism remains to be discovered. The most likely mechanisms, on the basis of current evidence, are shown in (Fig. 5.24) and microbes may use one or a combination of these pathways. The remainder of the molecule (alkyl side chain and benzene ring) is the same as the organic component of the alkyl benzene sulphonates and is presumably metabolized in a similar way.

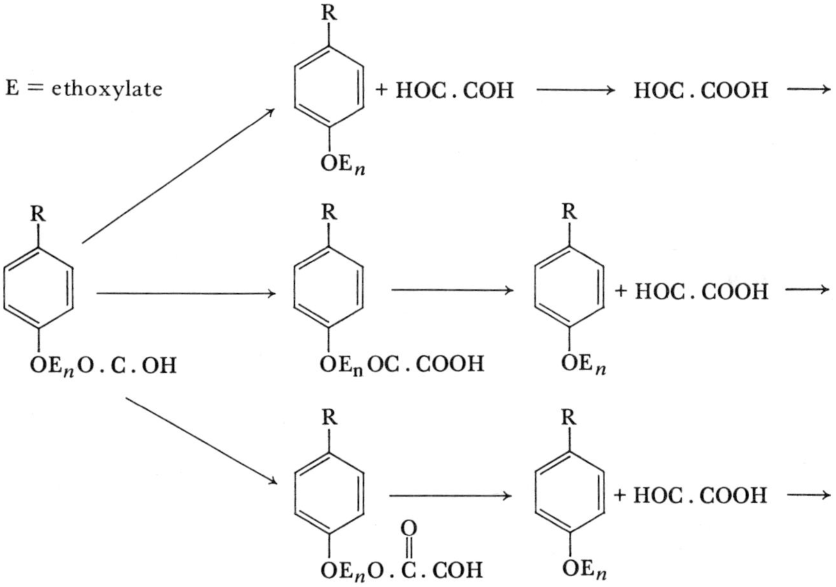

E = ethoxylate

R, OE_n + HOC . COH ⟶ HOC . COOH ⟶

R, OE_nO . C . OH

R, OE_nOC . COOH ⟶ R, OE_n + HOC . COOH ⟶

R, OE_nO . C(=O) . COH ⟶ R, OE_n + HOC . COOH ⟶

Fig. 5.24. Alkylphenol ethoxylate metabolism by microbes.

3. Cationic Surfactant Degradation

It is clear that microbes are capable of metabolizing cationic surfactants but details of the biochemistry are unavailable. The long hydrophobic alkyl side chains are presumably degraded in the usual way, i.e. by terminal oxidation followed by β-oxidation to acetate units. Some cationics contain pyridine rings and may be degraded in an analogous way to the herbicides, paraquat and diquat (Chapter 2).

ECOLOGICAL CONSIDERATIONS

1. Effect of Surfactants on Microorganisms

The relationship between surfactants and microbes is complex and involves factors other than biodegradation. Under appropriate conditions surfactants can act as bacteriocides or may simply be adsorbed by the microorganism. It is the ionic surfactants that are toxic to bacteria. The toxicity of the cationics increases with increasing pH whilst that of the anionics increases with decreasing pH reflecting the dissociation of the molecule. At neutral pH the cationics are the most toxic. In addition to pH the effectiveness of a particular surfactant as a bacteriocidal agent depends upon a number of factors, such as the particular microbial species, the size of the hydrophobic portion of the molecule, the presence of organic matter and the presence of divalent metal ions. Two of these factors are considered in detail below.

1.1. Species Effects

Cationic surfactants are equally effective inhibitors of both Gram-positive and Gram-negative microorganisms, whilst Gram-positive species are most susceptible to the anionics. Many bacteria are entirely unaffected by Teepol (a secondary alkyl sulphate) at 500 ppm but others (e.g. *Staphylococcus aureus*) are sensitive to concentrations as low as 200 ppm. TBS inhibits the respiration of both this bacterium and of *Mycobacterium phlei* at 10 ppm and will kill them at 1000 ppm. *Proteus vulgaris* is more resistant, the corresponding figures for death being 1000 and 10 000 ppm. On the other hand, some species have been shown to survive and even grow at very much higher concentrations (250 000 ppm). This, however, is unlikely to be a simple species effect, and may be dependent upon additional factors such as the presence of other nutrients and perhaps micelle formation, which would reduce the effective concentration of the surfactant.

The difference in sensitivity to the anionic surfactants, between Gram-positive and Gram-negative species, is several orders of magnitude. Many Gram-positives are noticeably affected by 10-20 ppm whilst several thousand ppm may be without effect on a Gram-negative organism. Strain differences in sensitivity are also common, for example one *Escherichia coli* strain shows 100% mortality at a particular concentration of SDS whilst another strain shows only 10% mortality.

1.2. Chemical Structure of the Surfactant

Anionic surfactants become increasingly toxic with increases in chain length. In many cases compounds containing a benzene ring seem to be less toxic than simple alkyl analogues. For example, kerylbenzene sulphonate is generally not as toxic to bacteria as sodium lauryl sulphate.

The swarming ability of the genus *Proteus* is inhibited by surfactants and with LAS there is a quantitative relationship between inhibitory activities and chain length. For example, the motility of *Proteus mirabilis* is 50% inhibited by 4000 ppm of the C_6 compound whilst the C_{14} homologue causes the same effect at 20 ppm.

1.3. Mechanisms of Surfactant Toxicity

The mechanisms of surfactant toxicity are at present poorly understood but probably the most important characteristic is their interaction with bacterial proteins and lipids. The effect of surfactants on restricting motility of *Proteus* species is due to the absence of flagella but whether this is due to some direct effect on the flagella or inhibition of their biosynthesis is not known. Many studies have been performed with sodium dodecyl sulphate, a primary alkyl sulphate which causes the disintegration of flagella, probably solubilizes membrane proteins and induces cell wall damage.

Most investigations of surfactant-protein interactions have used purified proteins or crude cell-free preparations and anionic surfactants but it is not clear how relevant results are to *in vivo* situations. These interactions may be dependent upon pH influenced ionic forces between the charged groups of the surfactant and those of the protein; in addition to non-ionic attractions between the hydrophobic portions of the molecules. The overall effect will vary with different proteins and different surfactants, but typically involves protein solubilization, protein precipitation, and association or dissociation of subunits. Both cationic and anionic surfactants behave similarly in that it is the hydrophobic forces that are the most important. An apparent anomaly is that the small amount of work that has been done with the non-ionics shows that these compounds have little effect on bacteria and do not interact readily with proteins. However, it may be that in the case of the ionic surfactants it is the initial ionic attraction that facilitates the more important hydrophobic interactions.

Intact surfactant molecules, if they should enter the bacterial cell, can have a considerable influence upon enzyme activities as demonstrated by the complex effects of surfactants on purified enzyme preparations. Many enzymes are inhibited by anionic surfactant concentrations in the region of 100-1000 ppm, an effect which tends to increase with any increase in the size of the hydrophobic portion of the molecule. For example, secondary octane sulphonate causes 50% inhibition of phosphatase, alkaline phosphatase, papain and invertase activity at 25 000-50 000 ppm whilst the C_{17} analogue has the same effect at much lower concentrations (25-600 ppm). These interactions are, in some cases, highly pH dependent, as illustrated by sodium dodecyl sulphate (SDS) which, at 300 ppm, completely inhibits urease at pH 5.0 but at pH 5.4 has no effect whatsoever. The cationic surfactants are, in general, less inhibitory towards enzymes than the anionics although the former are more toxic towards intact bacteria. In fact, moderate concentrations of cationics often have an apparently anomolous affect on enzymes involved in fat metabolism, increasing rates, sometimes by factors as high as ten. This has been attributed to the emulsification of fatty substrates by the surfactant.

2. Mechanisms of Enzyme Inhibition by Surfactants

There are very few experiments pertinent to the mechanisms of enzyme inhibition by surfactants. However, those that are suggest that even though they may not be the only factors involved, ionic interactions are of fundamental importance.

For instance, the charged terminal amino groups of lysine residues, in the proteolytic enzyme trypsin, can be neutralized by acetylation or succinylation whilst still retaining most of their catalytic function. The reduction of surfactant inhibition following this treatment is thought to be due to a decrease in ionic interaction between the now modified enzyme and the surfactant. SDS, at a concentration of 300 ppm, does not affect these derivatives but still causes an 88%

inhibition of the native enzyme. This is because the anionic surfactant binds to the positive amino groups in the natural protein but not to the neutralized groups in the derivatives. Similarly cationic surfactants are without effect on the natural protein but do inhibit the derivatives; here neutralization of the positively charged groups on the protein facilitates access to the negatively charged ones. Non-ionic surfactants affect neither the native enzyme nor the derivatives. Some ionic surfactants have been shown to increase the intracellular catalase activity of the yeast *Candida albicans.*

3. Adsorption of Surfactants by Bacteria

Many microbes adsorb ionic surfactants quite strongly and this is a most important consideration in the interpretation of biodegradation data. Apparent disappearance of a surfactant may not necessarily reflect biodegradation but may be primarily due to adsorption. Anionics are the most strongly adsorbed and this is presumably a reflection of the surfactant-protein interactions discussed above. The adsorption phenomenon is obviously complex involving many different factors (Chapter 2), and is usually dependent upon pH and metal ions. For example, the adsorption of TBS is increased several fold in the cases of *Escherichia coli* and *Staphylococcus aureus* by the presence of 1000 ppm of calcium.

4. The Use of Surfactants in the Pesticide Industry

Surfactants are used as detergents, emulsifiers and wetting agents in the pesticide industry. The addition of surface active agents to herbicide formulations can have dramatic effects on their subsequent toxicity. For example, 500 ppm of picloram + 1% non-ionic surfactant is as potent as 2000 ppm of picloram alone. This synergism may be due to increased penetration of the herbicidal fraction although, in some cases, the surfactant itself has inherent toxicity. The individual and combined contribution of the surfactant varies according to the concentration and type of compound used. The non-ionic Tween 80 (a derivative of sorbitan monooleate) has been shown to cause elongation of wheat coleoptile sections at concentrations from 0.001 to 0.1 $\mu g\ ml^{-1}$. Tween 40 (a derivative of sorbitan monopalmitate) produces no response whilst Tween 20 (a derivative of sorbitan monolaurate) inhibited growth. The variability in the fatty acid component of the molecule is suggested as responsible for this effect on cell elongation. A range of surfactants have also been demonstrated as inhibitors of cell division in pea root meristems. Triton X-100 (a non-ionic octyl phenol surfactant) may inhibit growth by uncoupling photophosphorylation; it also enhances phosphorus absorption and depresses magnesium uptake.

Structural modifications to the surfactant may greatly improve the efficiency of the herbicide with which it is mixed and in a way that cannot be explained by a simple summation of individual effects. The activity of amitrole, dalapon, 2,4-D, and dinoseb can be increased when combined with straight chain isomers of the anionic dodecylbenzene sulphonate, an increase that becomes more apparent as the position of the benzene ring is moved towards the centre of the 12-carbon *n*-dodecyl alkyl group. Other structural changes, which induce a stimulatory effect on herbicidal activity, include the presence of an highly branched dodecyl alkyl group and the substitution of an ester oxygen atom for the benzene ring.

Non-ionic surfactants have also achieved prominence in the herbicide industry because of their greater compatibility when formulated in water with high salt content. Since non-ionic surfactants contain both hydrophobe and hydrophile moieties they can be variously synthesized to achieve specific solubility characteristics. One of this group—the alkylphenol polyoxyethylene glycol esters—has been evaluated for structure/activity relationships. These surfactants consist of an alkylphenol hydrophobe and ethylene oxide (EO) hydrophile. The number of moles of EO in the molecule determines its surface tension, solubility and hydrophilic-lipophilic balance (HLB). When used in combination with amitrole, dalapon, or paraquat as a foliar spray of maize, toxicity reaches a maximum at 8-15 moles of EO above which it decreases. The capacity of a surfactant to enhance foliar penetration is a function of its HLB together with other chemical and physical properties. HLB is expressed numerically and ranges from lipophilic values of two, through a neutral range of about ten, to extremely hydrophilic values of greater than thirty.

In the light of the above information it has been suggested that the herbicide and surfactant might be incorporated into the same molecule. As a result, long chain alkylamine salts of 2,4-D have been prepared and are being evaluated.

5. Role of Detergent Preparations in Eutrophication

Detergent preparations make a significant contribution to phosphate in natural waters although it is clear that this is only one of many factors contributing to eutrophication. It is, in fact, difficult to assess accurately the amount of detergent-derived phosphate in rivers, streams and lakes. It appears, however, that in populated areas approximately 50% of the phosphorus comes from land drainage (including farm wastes and general agricultural run-off) whilst the remainder comes from sewage and industrial effluents. The amount of ortho-phosphate in sewage that comes from detergents varies between 32 and 70%, so that detergent phosphate represents between 16 and 35% of the total phosphate in natural waters. Since in most waters the phosphate content is considerably in excess of the threshold level for algal growth and since it has been shown that algal growth is not proportional to phosphate concentration above this level, it is highly unlikely that even the complete removal of phosphates from detergents would have any appreciable effect on eutrophication, unless many other factors are also modified.

6. Phosphate-Free Detergents

The demand for phosphate-free detergents, in North America, has led to a search for an alternative builder with the combined functions of polyphosphates i.e. to chelate interferring metal ions, aid the breakdown of fats and oils and to prevent deposition of suspended dirt from the washing water. It is not easy to find such a substitute that is both non-toxic and readily biodegradable. Polyphosphate itself is biodegraded extremely rapidly and is usually hydrolysed to orthophosphate by the time it reaches a sewage plant. One alternative, widely used in detergents especially in North America, is trisodium nitrilotriacetate or NTA ($Na_3N(CH_2COO)_3$). In laboratory studies it is broken down quite readily, although in a trickling filter sewage system NTA may be only 60-80% biodegraded prior to discharge. Although NTA is itself non-toxic there is a danger that the partially degraded material could give rise to nitrosamines that are known carcinogens and are clearly highly undesirable, particularly if they find their way into drinking water. Aerobic and anaerobic microorganisms have been isolated that degrade NTA and the detailed route by which this compound is metabolised by a *Pseudomonas* species is shown in Fig. 5.25.

7. Detergent Additives

Commercial detergent preparations usually contain substances other than the surfactant proper. These additives have a variety of functions but in the broadest sense they increase the effectiveness of

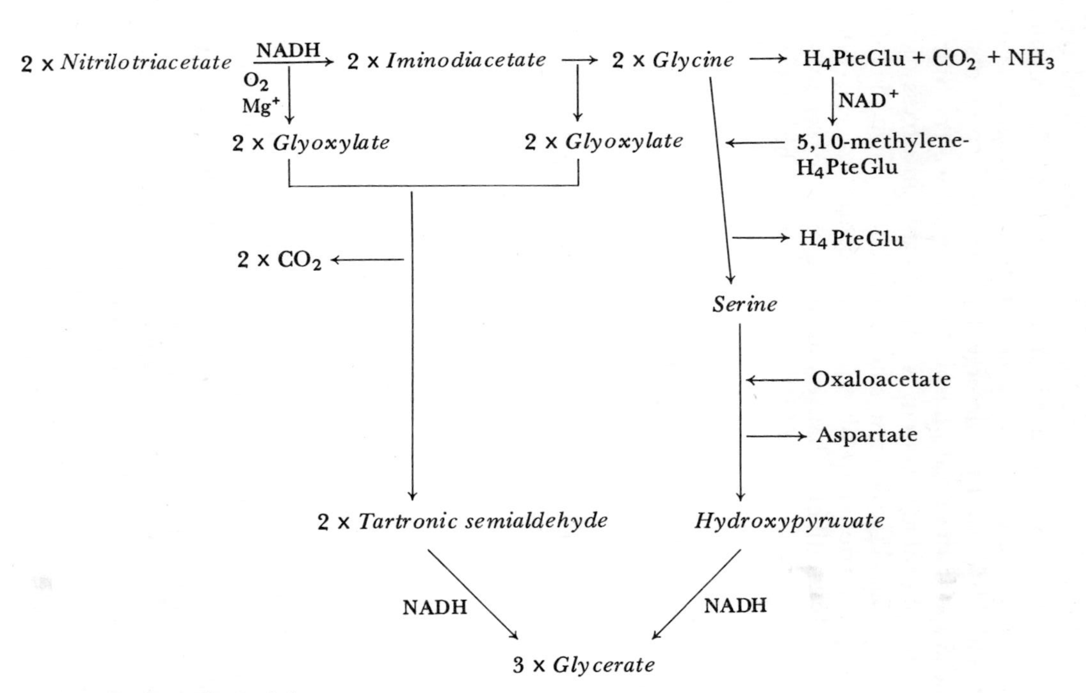

H_4PteGlu = tetrahydrofolate coenzyme

Fig. 5.25. Metabolism of nitrilotriacetate by a Pseudomonad.

the surfactant in the special conditions for which it is formulated. It seems appropriate to consider these substances in this chapter although they are not themselves surfactants.

The formulation of a typical domestic preparation is shown in Table 5.1. The major component is named the builder and for a long time polyphosphate salts have filled this role. In addition, optical

TABLE 5.1

Typical percentage compositions of household washing products

Constituent	Heavy duty powders (USA)	Heavy duty powders (Europe)	Light duty powders
Surface active ingredient	14–20	14–18	25–32
Foam booster	1.5–2	1–3	2–15
Sodium tripolyphosphate	40–60	30–45	2–15
Optical brighteners, minor ingredients, sodium sulphate, water	up to 100		

brighteners and perborates are added in smaller amounts. In recent years there has been much argument about the polluting effects of polyphosphate builders, the most widely used of these being sodium tripolyphosphate. It has been claimed that these phosphates are in part responsible for eutrophication of lakes (Chapter 3), particularly in the U.S.A., Canada and Sweden. This has led, particularly in North America, to a considerable amount of public and legislative pressure aimed at reducing phosphate in, or even eliminating it from, commercial preparations. All this has happened, even though the case against detergent phosphates is unproven and indeed there are currently moves to reverse or modify some of the decisions. Phosphates, of course, cannot be considered to be poisonous or directly harmful to man. In addition, the entry of phosphates from detergents into inland waters may play only a small part in the eutrophication process.

CONCLUSIONS

The majority of surfactants and additives used in commercial preparations appear non-harmful to man and other animals. Most present day synthetic surfactants are non-persistent and are degraded primarily by microorganisms. Some of the early synthetic surfactants were highly branched, biodegraded only rather slowly and likely to cause foaming. Their replacement with unbranched analogues has largely overcome these problems.

There remains the controversy over the adverse effects on inland waters of the polyphosphate builders and the possible health hazards of alternative additives.

Recommended Reading

Cain, R. B. (1975). Surfactant biodegradation. *In* "Industrial Effluent Treatment" (A. G. Callely, C. F. Forster and D. Stafford, Eds.). English University Press, London.

Gledhill, W. E. (1974). Linear alkylbenzene sulfonate: biodegradation and aquatic interactions. *Advances in Applied Microbiology* 17, 265.

Jungermann, E. (Ed.) (1970). "Cationic Surfactants." Marcel Dekker, New York.

Swisher, R. D. (1970). "Surfactant Biodegradation." Marcel Dekker, New York.

6
Synthetic Polymers

INTRODUCTION

The dramatic development of the petrochemical industry during the last thirty or forty years has led to the large scale production of three main groups of polymeric materials: plastics (e.g. PVC, polyethylene, polypropylene, PTFE), synthetic elastomers or rubbers (e.g. styrene-butadiene rubber, neoprene rubber) and "man-made fibres" (e.g. polyamide, polyester, polyacrylonitrile). These substances have made major contributions (mostly beneficial) to our way of life although some cynics have suggested that engulfment in synthetic polymers may be a fitting end to this technological era.

One obvious indication of the demand for synthetic polymers is America's seven fold increase in plastic production between the years 1950 and 1968, a trend which is continuing (Table 6.1). The world growth rate for total synthetic polymers is still over 10% per annum.

TABLE 6.1
Synthetic polymer production (tonnes x 10^{-6})

Polymer	Region	1965	1968	1970
Polyamide	U.S.A.	0.53	0.66	0.85
Polyester	U.S.A.	0.17	0.50	0.75
Polyacrylonitrile	U.S.A.	0.15	0.26	0.31
Total synthetic polymers	World	2.0	3.75	4.7

Plastics, rubbers and "man-made fibres" have a wide range of physical properties making them useful in a variety of roles. This versatility, combined with cheapness, ease of production and resistance to decay, are the main reasons for their widespread adoption. However, it is this very persistence that has given cause for concern.

The visual pollution due to synthetic polymers, especially plastics, must be obvious to anyone visiting a beach in most parts of the world. Yet despite the aesthetic displeasure that synthetic polymers cause they do not appear to pose a serious biological threat. They are neither toxic nor water soluble, do not enter food chains and, when eventually degraded, the products are usually harmless.

CHEMISTRY

1. Plastics

Plastics are synthetic polymers that have stable properties when in normal use but which, at some stage in their manufacture, can be shaped or moulded by heat or pressure or both.

1.1. Plastics Derived From Polyalkenes

It is a characteristic of vinyl compounds that they undergo addition polymerizations to form a variety of macromolecules (Fig. 6.1) whose properties are dependent upon the conditions of the reaction and the individual monomer.

$$n\mathrm{CH_2{=}CH.X} \longrightarrow -\mathrm{CH_2\underset{\displaystyle X}{\underset{|}{C}}H\ CH_2{=}\underset{\displaystyle X}{\underset{|}{C}}H\ CH_2{=}\underset{\displaystyle X}{\underset{|}{C}}H}-\ \mathrm{etc}$$

Fig. 6.1. Polymerization of alkenes.

The most important plastics derived from polyalkenes are shown in Table 6.2 whilst their formulae appear in Fig. 6.2.

TABLE 6.2
World production of polyalkene-derived plastics (tonnes x 10^{-6})

Polyethylene	8
Polystyrene	>3
Polypropylene	~1
Polyvinylchloride	>3
Polyacrylonitrile	>1
Polyvinylacetate	~1

Polyethylene

$$-(CH_2-CH_2)_n-CH_2CH_2CH_2CH_2-(CH_2-CH_2)_n-$$

Polypropylene

$$-(CH_2-CH_2-CH_2)_n-CH_2-CH_2-CH_2-(CH_2-CH_2-CH_2)_n-$$

Polystyrene

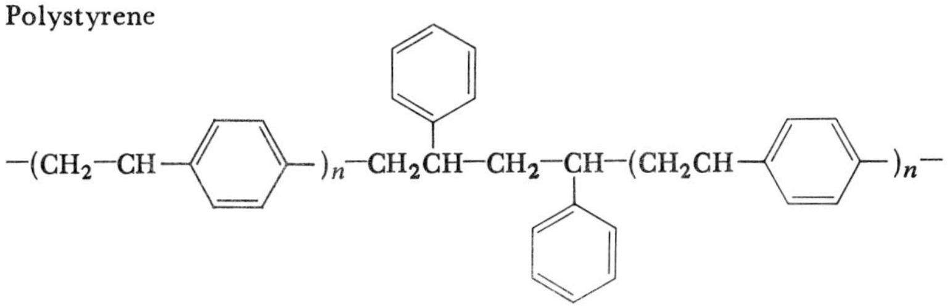

$$-(CH_2-CH-C_6H_4-)_n-CH_2CH(C_6H_5)-CH_2-CH(C_6H_5)-(CH_2CH-C_6H_4-)_n-$$

Polyvinylchloride (PVC)

$$-(CH_2-CHCl)_n-CH_2-CHCl-CH_2-CHCl-(CH_2-CHCl)_n-$$

Polyacrylonitrile

$$-(CH_2-CHCN)_n-CH_2-CHCN-CH_2CHCN-(CH_2-CHCN)_n-$$

Polyvinylacetate

$$-(CH_2-CHOOC-CH_3)_n-CH_2-(CHOOC-$$
$$CH_3)-CH_2-(CHOOC-CH_3)-(CH_2-CHOOC-CH_3)_n-$$

Fig. 6.2. Formulae of polyalkenes.

The vinyl chloride for PVC synthesis is prepared from ethylene by the method shown in Fig. 6.3. Vinyl acetate is also produced from ethylene by catalytic condensation with acetic acid (Fig. 6.4).

$$CH_2{=}CH_2 + Cl_2 \longrightarrow \underset{\textit{ethylene dichloride}}{CH_2Cl-CH_2Cl} \xrightarrow[-HCl]{} \underset{\textit{vinyl chloride}}{CH_2{=}CHCl}$$

Fig. 6.3. Synthesis of vinyl chloride.

$$CH_2{=}CH_2 + CH_3COOH \xrightarrow[O_2]{\text{Pd/Cu catalyst}} \underset{\textit{vinyl acetate}}{CH_2{=}CHOOC-CH_3} + H_2O$$

Fig. 6.4. Synthesis of vinyl acetate.

The monomer for polystyrene polymers is prepared by the direct ethylation of benzene (Fig. 6.5).

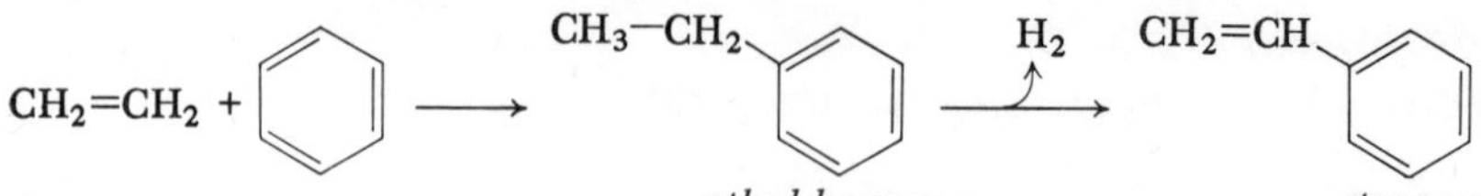

$$CH_2{=}CH_2 + C_6H_6 \longrightarrow CH_3{-}CH_2{-}C_6H_5 \xrightarrow{H_2\uparrow} CH_2{=}CH{-}C_6H_5$$

ethyl benzene *styrene*

Fig. 6.5. Synthesis of styrene.

1.2. Other Plastics

Although the majority of modern plastics are alkene-derived, there are other types produced in large quantities, the best known being a rather complex polymer, polyurethane. Fig. 6.6 represents part of a polyurethane chain but many alternative structures exist.

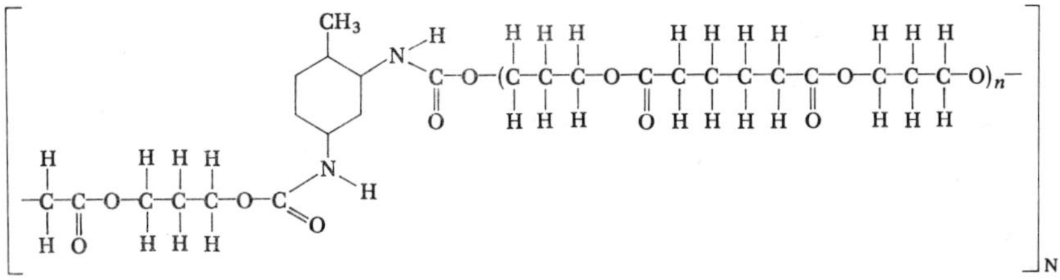

Fig. 6.6. Structure of polyurethane.

2. Elastomers

Elastomers are polymers with resilient, elastic properties. The naturally occurring member of the group is rubber for which there has been an immense increase in use in the twentieth century. World production in 1900 was 50 000 tonnes and this figure had increased to 2.5 million by 1970. Notwithstanding the production of synthetic rubber is now approximately twice that of natural rubber. The major precursors of synthetic elastomers are butadiene and isoprene and in fact some 90% of world butadiene is used in their production.

The properties required of an elastomer hinge very much upon the particular purpose for which it is intended. For example, if thermal extremes are likely to be encountered the product should be stable over a wide temperature range or if there is a fire hazard should be non-flammable. Resistance to oils, solvents and abrasives is also important under certain circumstances. Some synthetic elastomers are better equipped to deal with one or more of these conditions when compared with natural rubber.

2.1. Styrene-Butadiene Rubber

Styrene-butadiene rubber is prepared by a free radical polymerization of a mixture of butadiene and styrene. The end-product has only a mediocre resistance to oils and solvents but is highly elastic.

2.2. Polybutadiene

Polybutadiene (Fig. 6.7) has good abrasion resistance and remains flexible at low temperatures.

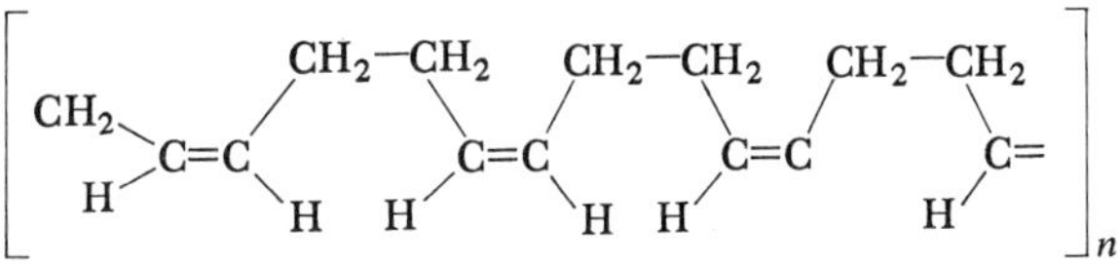

$$\left[\begin{array}{c} CH_2 \\ H \end{array} \!\!>\! C{=}C \!<\! \begin{array}{c} CH_2{-}CH_2 \\ H \quad H \end{array} \!>\! C{=}C \!<\! \begin{array}{c} CH_2{-}CH_2 \\ H \quad H \end{array} \!>\! C{=}C \!<\! \begin{array}{c} CH_2{-}CH_2 \\ H \end{array} \!>\! C{=} \right]_n$$

Fig. 6.7. Structure of polybutadiene.

2.3. Neoprene Rubber

Neoprene rubber is prepared by the catalytic polymerization of chloroprene [I]. It is expensive, yet tolerant of weather extremes, does not burn and is little affected by hydrocarbons.

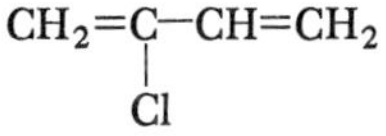

$$CH_2{=}\underset{\displaystyle Cl}{\underset{|}{C}}{-}CH{=}CH_2 \qquad [I]$$

2.4. Butyl Rubber

Butyl rubber is a co-polymer of isobutene [II] and isoprene in the ratio of about 19:1. Since it is highly saturated, butyl rubber is resistant to chemical attack, resilient and impermeable to gases. It is used for inner tubes and for the internal lining of tubeless tyres.

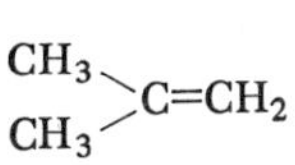

$$\begin{array}{c} CH_3 \\ CH_3 \end{array} \!\!>\! C{=}CH_2 \qquad [II]$$

2.5. Synthetic Polyisoprene

Synthetic polyisoprene (Fig. 6.8) is an analogue of natural rubber. The main advantages over its natural counterpart lie in production uniformity and, until the recent oil crisis, price stability.

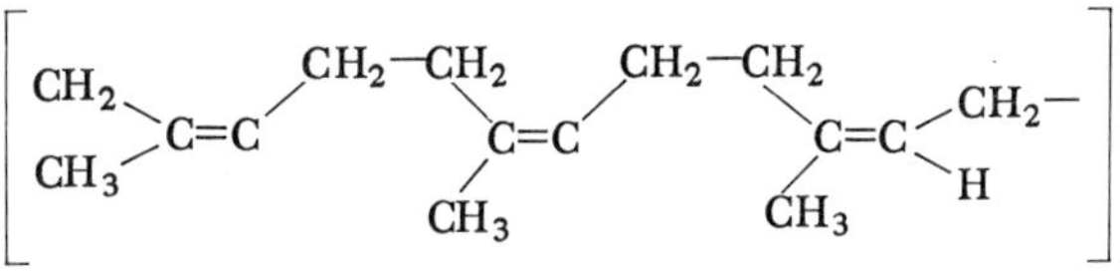

$$\left[\begin{array}{c} CH_2 \\ CH_3 \end{array} \!\!>\! C{=}C \!<\! \begin{array}{c} CH_2{-}CH_2 \\ CH_3 \end{array} \!>\! C{=}C \!<\! \begin{array}{c} CH_2{-}CH_2 \\ CH_3 \end{array} \!>\! C{=}C \!<\! \begin{array}{c} CH_2{-} \\ H \end{array} \right]_n$$

Fig. 6.8. Structure of polyisoprene.

2.6. Nitrile Rubber

The nitrile rubbers are co-polymers of about 30% acrylonitrile and a diene (usually butadiene). They have good mechanical properties, are resistant to heat and oil and are cheap to produce.

2.7. Ethylene-Propylene Elastomers

Ethylene-propylene elastomers (EP rubbers) are synthesized by the co-polymerization of ethylene with propylene over co-ordination catalysts (e.g. vanadium oxychloride, ethyl aluminium chloride). These are relatively recently developed elastomers but their use is increasing and annual world production is at present about 100 000 tonnes. These compounds are resistant to oxidation, discolouration and heat.

3. Synthetic Fibres

Synthetic fibres are largely derived from aromatic and alicyclic hydrocarbons.

3.1. Polyamides

Cyclohexane is a key intermediate in the synthesis of polyamides (Nylon 6, Nylon 6,6). The manufacture of Nylon 6 is summarized in Fig. 6.9 and that of Nylon 6,6 in Fig. 6.10.

NO
$h\nu$
NOCl
H_2SO_4
cyclohexane
nitrosocyclohexane
CO
NH
O
NOH
caprolactam
NH_2OH
cyclohexanone
cyclohexanone oxime
high temperature

$[NH.(CH_2)_3.CO.(CH_2)_5.CO]_x$

NYLON 6

Fig. 6.9. Manufacture of Nylon 6.

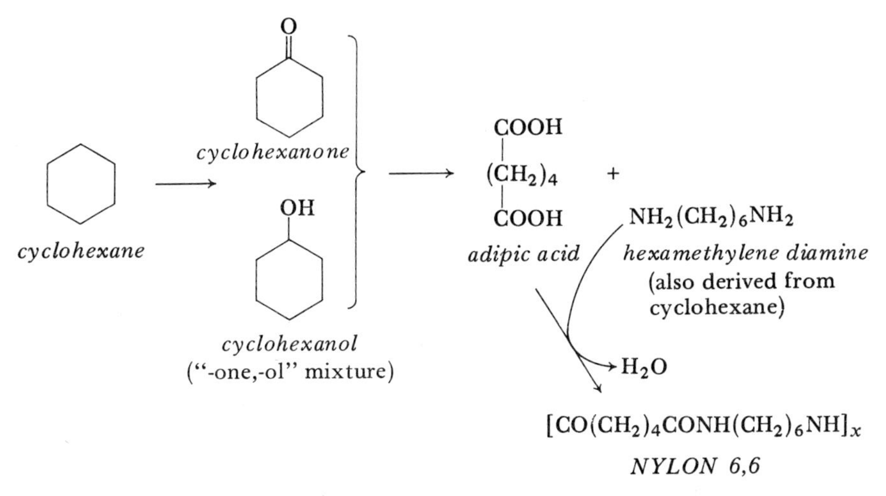

Fig. 6.10. Manufacture of Nylon 6,6.

3.2. Polyesters

Polyester fibre is prepared by polymerizing terephthalic acid (obtained by oxidation of *p*-xylene) with ethylene glycol (Fig. 6.11). The polymer is spun into fibres and given a variety of trade names such as Terylene and Dacron. Annual production figures for polyesters are on the increase.

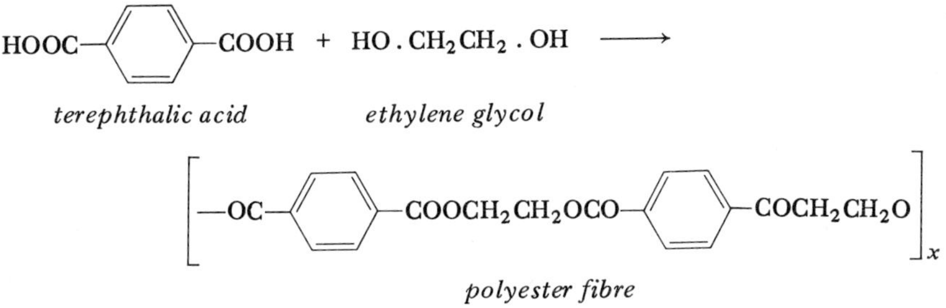

$$HOOC-C_6H_4-COOH + HO.CH_2CH_2.OH \longrightarrow$$

terephthalic acid *ethylene glycol*

$$\left[-OC-C_6H_4-COOCH_2CH_2OCO-C_6H_4-COCH_2CH_2O\right]_x$$

polyester fibre

Fig. 6.11. Synthesis of polyester fibre.

3.3. Polyacrylonitriles

Polyacrylonitrile is prepared by the catalytic polymerization of its monomer, acrylonitrile [III]. The polymer has many wool-like properties but is relatively poor wearing.

$$CH_2{=}CH{-}CN \qquad [III]$$

ANALYTICAL METHODS

Studies concerning the effects of environment on synthetic polymers fall into two major areas: measurement of non-microbiological alterations in physical and chemical properties (weathering trials) and tests for microbial decomposition.

1. Weathering Trials

Samples are exposed for varying periods of time and any changes determined, commonly by the onset of embrittlement which can be measured in various ways:

A. A thick film of the material is slowly stretched, increasing the force gradually until it breaks. The elongation at the time of breaking is then measured.

B. Standard falling weights strike the plane of the polymer film until the force required to rupture it is determined. This property is described as the falling weight impact strength.

C. Simple flexing of the film through 180° by hand.

D. A decrease in the number average molecular weight often indicates embrittlement as do density changes.

E. Infra-red spectroscopy can be used to measure the numbers of carbonyl and vinyl groups in the polymer which may decrease during the weathering process.

2. Microbial Decomposition

2.1. Embrittlement

In the case of plastics, microbes are often involved in the degradation of plasticizers (additives which render the plastic more flexible) causing embrittlement. Many of the tests described above are used to measure the involvement of microbes in this process.

2.2. Biodegradation

Various methods are used in the study of polymer biodegradation. For example, thin films of the polymer or polymer formulation are deposited on microscope slides which are placed in suitable environments (e.g. soil, water) and examined after an appropriate time for microbial colonization. Any microbes observed may then be isolated and tested in either pure or mixed culture for their ability to degrade or modify the polymer or any of its additives. If the polymer is supplied as the sole carbon and/or nitrogen source to a microbial culture, biodegradation may be measured by ammonia release, oxygen utilization, carbon dioxide production or weight loss. If decay in soil is examined it is often difficult to obtain meaningful results because microbial activity is extremely variable and the rates of synthetic polymer metabolism are usually very low. The sensitivity of biodegradation measurements can be greatly enhanced by using carbon-14 labelled polymers and measuring carbon-14, either in assimilated organic matter or released in carbon dioxide.

SYNTHETIC POLYMER DECAY

Synthetic polymers are subjected to several degradative factors in the environment, the most important of these being oxygen, rainfall, UV and visible radiation and biological effects. Most polymers are stable and there are many reports of them remaining essentially unchanged after several years in the soil or in the sea. The recalcitrance of these compounds is probably due to one or a combination of the following: (i) molecular structures in which the main chains of the molecule are composed of carbon-carbon bonds requiring large amounts of energy for cleavage; (ii) poor solubility in water and lipids; (iii) high molecular weight; (iv) highly branched structure.

Admittedly many microorganisms metabolize fatty acids and hydrocarbons, which, like the synthetic polymers, are mainly composed of chains of directly linked carbon atoms. Obviously this characteristic alone is not sufficient to prevent decay. However, those fatty acids and hydrocarbons that are also highly branched are somewhat resistant to degradation, even though their water solubility is low and they are soluble in microbial lipid membranes where the enzymes involved in their breakdown are located. Linear, high molecular weight hydrocarbons (those with more than thirty carbon atoms) are much less soluble in lipids than are their lower homologues and are correspondingly more persistent. Natural high molecular weight polymers that are readily degraded biologically have, with few exceptions, chains of mixed linkages and the bonds attacked (often by hydrolytic enzymes) are often those between the carbon and another atom (usually oxygen or nitrogen). It should be emphasized that these are broad generalizations concerning the decay of synthetic and natural polymers. For instance, Nylon contains mixed linkages similar to those in proteins, yet is not subject to microbiological attack whilst natural rubber, containing mainly carbon-carbon bonds, is slowly degraded.

An important distinction, particularly in a discussion of plastic decay is that between disintegration and degradation. Disintegration without degradation can be brought about by physical forces (e.g. abrasion by sand) or biological ones (burrowing and gnawing activities of soil and marine animals). In addition, some plastics contain a plasticizer which is relatively easily metabolized by soil microbes resulting in the disintegration of the polymeric material even though the chemical nature of the polymer itself has not been altered. PVC plastics may contain up to 50% plasticizer and, when this additive is a sebaceate about 40% of it will have been degraded after fourteen days in the soil.

1. Non-Biological

In general, elastomers are susceptible to oxidation since they contain double bonds. Many plastics are also subject to oxidation and slow depolymerization when exposed to UV radiation. Indeed, in many cases, the recalcitrance of plastics is partially due to the addition of antioxidant "screening compounds" to prevent deterioration during use. Antioxidants are also important during plastic production when high temperatures are involved.

The oxidation process results from the presence of sensitive functional groups, either as integral components of the polymer chain or as impurities. The most important of these groups is hydroperoxide which, under conditions where it is thermally stable, is photolyzed by UV light to free radicals which then initiate the oxidation process (Fig. 6.12).

This means that hydroperoxides are powerful photoactivators of autooxidation and their presence greatly reduces the period required for degradation. In contrast, some polymers (polyesters, polyurethanes and cellulose esters) are embrittled and slowly degraded primarily by hydrolytic mechanisms.

$$\mathrm{ROOH} \xrightarrow{\mathrm{UV}} \mathrm{RO\cdot + \cdot OH}$$

$$\mathrm{RO\cdot + 2RH} \longrightarrow \mathrm{ROOH + ROH + R\cdot}$$

$$\mathrm{HO\cdot + 2RH} \longrightarrow \mathrm{ROOH + H_2O + R\cdot}$$

Fig. 612. Formation of free radicals from hydroperoxide.

2. Biological

Prior to 1939 it was believed that synthetic polymers were resistant to biological attack. However, during World War II, some plastics used in tropical regions were found to deteriorate quite rapidly. There were many studies of this phenomenon but it was not established whether the microbes involved were degrading additives, such as the plasticizers, or the polymers themselves. Indeed in the light of more careful, recent work (mainly with plastics) it is clear that microbial decomposition of the actual synthetic polymer is a relatively rare phenomenon although microbes do, by metabolizing additives, bring about major physical changes. There are various levels of biological interaction with plastic formulations.

2.1. Mechanical

Termites, insects, molluscs and rodents are responsible for the destruction of plastics and other synthetic polymers.

2.2. Biochemical

Microorganisms affect synthetic polymers in two main ways, Firstly, they have a limited capacity to metabolize these compounds and secondly microbial secretions may damage them. There are only a few cases where strong evidence suggests that microbes are involved in the degradation of commercially important polymers *per se* (e.g. cellulose nitrate, polyvinyl acetate). Therefore, probably the most widespread effect that microbes have on plastics involves the metabolism of plasticizers. This causes changes in physical properties and facilitates the subsequent degradation of the polymer by purely chemical processes. Most plasticizers are susceptible to microbial attack and the capacity of thermophilic soil microflora to metabolize a variety of them is indicated by the data in Table 6.3.

TABLE 6.3

The effect of addition of plasticizers on oxygen consumption by thermophiles

Plasticizer	Percentage change in respiration (over 72 hour period)
Tri-*n*-butyl citrate (TNBC)	+1390
Di-octyl sebacate (DOS)	+100
Polyethylene glycol 200(PEG200)	+91.5
Tri-cresyl phosphate (TCP)	+84.5
Tri-tolyl phosphate (TTP)	+59.0
Di-octyl phthalate (DOP)	+28.5

Fungi are especially adept at growing on plasticizers as their sole carbon and energy source and this, to some degree, reflects the capacity of the mycelia to penetrate the plastic. Polymers which are normally resistant to microbial attack, can be rendered biodegradable if they are first treated chemically to yield low molecular weight carboxylic acids. This mechanism could form the basis of a waste treatment process for recalcitrant synthetic polymers. Table 6.4 lists common synthetic polymers for which there is evidence for microbial degradation *per se*. It is important to emphasize that in most cases the rates of biodegradation are extremely slow, being measured in months or years rather than days. There is little known about the enzymology and metabolic pathways involved in these processes.

TABLE 6.4

Synthetic polymers subject to microbial decomposition

Cellulose nitrate
Cellulose acetate (cellophane)
Caprolactone polyester
Polyethylene succinate
Polyethylene adipate
Polytetramethylene succinate
Polyvinyl acetate
Styrene butadiene (and other styrene copolymers)
Butyl acrylonitrile
Butadiene acrylonitrile

ECOLOGICAL CONSIDERATIONS

1. Development of More Readily Degradable Plastics

It is only in recent years that the polymer chemist has turned his attention to the synthesis of short-lifetime plastics. Hitherto, one of his major concerns has been the development of stabilizers that enable plastics to withstand the high temperatures used in processing and to increase their longevity. This latter characteristic is still clearly desirable for a variety of uses such as building materials, aircraft and motor car components, plastic buckets, dust-bins, etc. However, plastic material used for packaging presents both a waste disposal and a litter problem and the ideal material would be stable during use but unstable when discarded.

Oxidative degradation tends to occur spontaneously especially at high temperatures and this involves a free radical process often

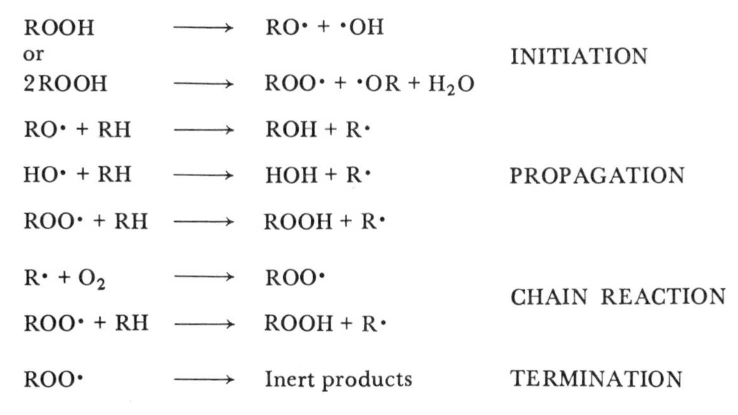

Fig. 6.13. Mechanism of auto-oxidation of polypropylene.

initiated by hydroperoxides (p. 180). Polypropylene, for example, is thought to be oxidized as shown in Fig. 6.13, a reaction which is prevented by the presence of antioxidant additives (usually phenolic compounds or amines) that function by removing the main chain-carrying species, the alkylperoxy radical (Fig. 6.14).

$$\mathrm{ROO^{\bullet} + AH \rightarrow ROOH + A^{\bullet}}$$

$$\mathrm{2A^{\bullet} \rightarrow non\text{-}radical\ products}$$

Fig. 6.14. Removal of alkylperoxy radicals by antioxidants.

Some transition metal dithiocarbamates act as peroxide decomposers (stabilizing agents) during polymer processing but on exposure to UV irradiation become activators of the auto-oxidation process. This phenomenon makes it feasible to produce a plastic that has a finite lifetime. Even the induction period, before photodegradation begins, can be controlled by using varying amounts of two additives, one a stabilizing ligand the other a metal ion activator.

Photodegradable plastics, based on this system, decompose to an hydrophilic powder which, once the molecular weight declines below 5000, may support microbial growth.

Three main ways are known of increasing the susceptibility of synthetic polymers to photodegradation and therefore the possibility of subsequent biodegradation.

1.1. Copolymerized Sensitizing Groups

A photosensitive co-monomer incorporated into the main chain or side chains of the polymer will give rise to UV initiated chain fission.

These sensitizing groups are usually carbonyl moieties which cleave as shown with the ethylene-carbon monoxide co-polymer in Fig. 6.15. After just a few days in sunlight, sheets containing 8% carbon monoxide, crumble on handling so that, even though complete degradation has not occurred, the material can no longer be described as litter.

$$CH_2{-}CH_2{-}CH_2{-}\underset{\underset{O}{\|}}{C}{-}CH_2{-}CH_2{\cdots}$$

$$\xrightarrow{h\nu}\ {-}CH_2{-}CH_2{-}CH_2{-}\underset{\underset{O}{\|}}{C}\ +\ {\cdot}CH_2{-}CH_2{\cdots}$$

$$\xrightarrow{h\nu}\ CH_3{-}\underset{\underset{O}{\|}}{C}{-}CH_2{-}CH_2{\cdots}\ +\ {-}CH{=}CH_2$$

Fig. 6.15. UV catalysed cleavage of ethylene-carbon monoxide co-polymers.

1.2. Photosensitizer Additives

A sensitizer or prodegradant may be included in the polymer and, following adsorption of radiation at a specific wavelength, acts as a source of free radicals (Fig. 6.16).

SENSITIZATION	S	$\xrightarrow{h\nu}$	R·
PROPAGATION	$R\cdot + O_2$	$\longrightarrow$	$RO_2\cdot$
	$RO_2\cdot + RH$	$\longrightarrow$	$RO_2H + R\cdot$
TERMINATION	$R\cdot + R\cdot$	$\longrightarrow$	R—R
	$R\cdot + RO_2\cdot$	$\longrightarrow$	RO_2R
	$2RO_2\cdot$	$\longrightarrow$	$RO_2R + O_2$

SCISSION

$${\cdots}\underset{\underset{H}{|}}{\overset{\overset{CH_3}{|}}{C}}{-}CH_2{-}\underset{\underset{\underset{\underset{H}{|}}{\underset{O}{|}}}{O}}{\overset{\overset{CH_3}{|}}{C}}{-}CH_2{-}\underset{\underset{H}{|}}{\overset{\overset{CH_3}{|}}{C}}{-}CH_2{\cdots}\longrightarrow HO\cdot + {-}\underset{\underset{H}{|}}{\overset{\overset{CH_3}{|}}{C}}{-}CH_2{-}\underset{\underset{O\cdot}{|}}{\overset{\overset{CH_3}{|}}{C}}{-}CH_2{-}\underset{\underset{H}{|}}{\overset{\overset{CH_2}{|}}{C}}{-}CH_2{\cdots}$$

$$\longrightarrow {\cdots}\underset{\underset{H}{|}}{\overset{\overset{CH_3}{|}}{C}}{-}CH_2\cdot + \underset{\underset{O}{\|}}{\overset{\overset{CH_3}{|}}{C}}{-}CH_2{-}\underset{\underset{H}{|}}{\overset{\overset{CH_3}{|}}{C}}{-}CH_2{\cdots}$$

Fig. 6.16. Mechanism of photosensitizer-initiated degradation.

There are two types of photosensitizers, transition metal and organic. Transition metal salts and complexes undergo electron transfer and adsorption of light at particular wavelengths giving rise to the free radical species. For instance, a ferric carboxylate will form an alkyl radical in the presence of UV light, (Fig. 6.17).

$$Fe^{3+}(^{-}OOC\text{—}R)_3 \longrightarrow Fe^{2+}(^{-}OOC\text{—}R)_2 + \cdot OOC\text{—}R \longrightarrow CO_2 + R\cdot$$

Fig. 6.17. Free radical formation from ferric carboxylate.

Various aromatic ketones and anthraquinones act as organic photosensitizers when incorporated into plastics at levels of around one percent. Benzophenone sensitizes polyethylene to both photodegradation and the initiation of the oxidative chain reaction described in Fig. 6.18.

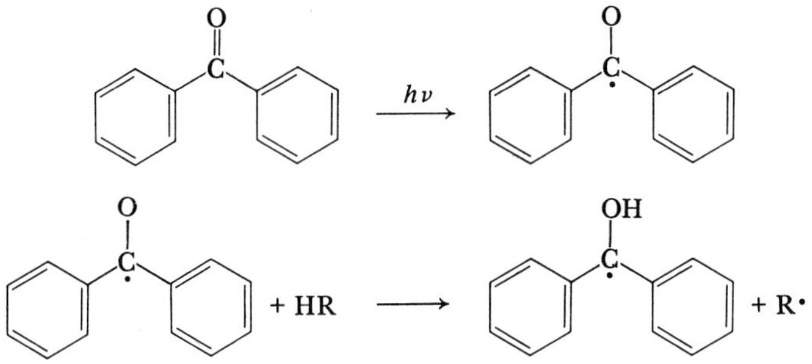

$$(C_6H_5)_2C{=}O \xrightarrow{h\nu} (C_6H_5)_2\dot{C}\text{—}O\cdot$$

$$(C_6H_5)_2\dot{C}\text{—}O\cdot + HR \longrightarrow (C_6H_5)_2\dot{C}\text{—}OH + R\cdot$$

Fig. 6.18. Benzophenone-initiated free radical formation.

1.3. Polymers of Low Photo-Mechanical Stability

1,2-Polybutadiene undergoes photocatalysed cross-linking and cyclization reactions which lead to the disintegration of the polymer (Fig. 6.19).

2. Practical Applications

For practical purposes synthetic polymers should be stable under the conditions of normal use, a characteristic which is often synonymous with a long indoor life. Polymers with carbonyl groups are especially suitable in this respect because they do not absorb above 330 nm and so are quite stable behind windows. There are a few polymers on the market that are degraded as a result of this principle. For example, a nylon has been developed (poly,1,3-phenylene isophthalimide, trade name Nomex), which is very susceptible to UV photodecomposition.

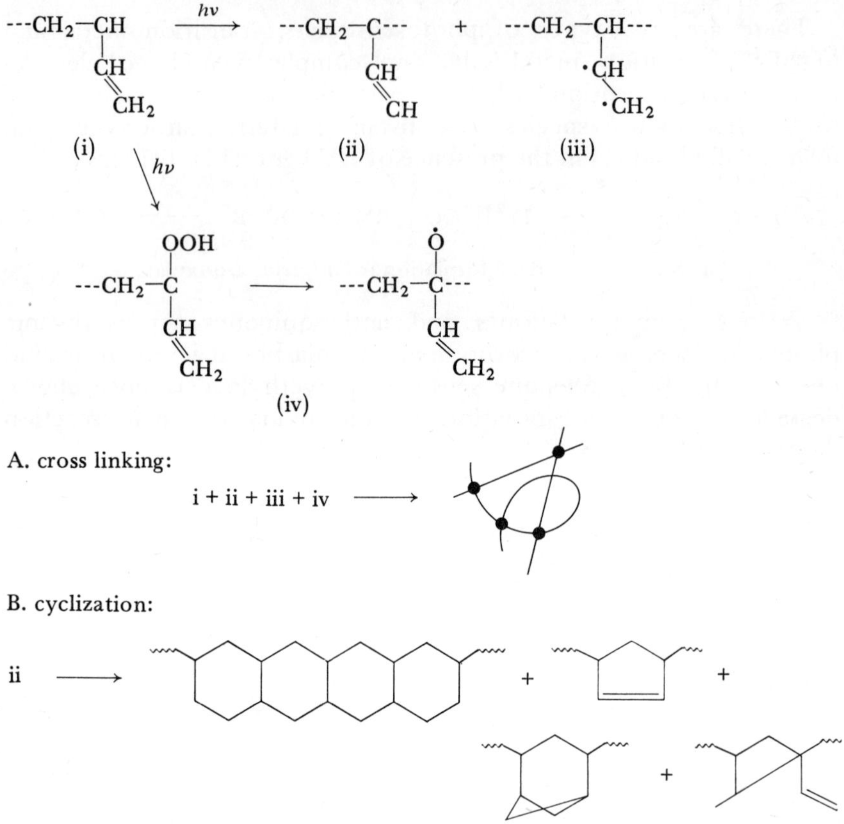

$$---CH_2-CH(-CH=CH_2) \xrightarrow{h\nu} ---CH_2-\dot{C}(-CH=\dot{C}H)--- + ---CH_2-CH(-\dot{C}H-\dot{C}H_2)---$$

(i) (ii) (iii)

$$\text{(i)} \xrightarrow{h\nu} ---CH_2-C(OOH)(-CH=CH_2) \longrightarrow ---CH_2-C(\dot{O})(-CH=CH_2)---$$

(iv)

A. cross linking:

i + ii + iii + iv ⟶

B. cyclization:

ii ⟶ + + +

Fig. 6.19. Photocatalysed degradation of 1,2-polybutadiene.

In addition, a few readily degradable plastics are now in production. One of these, Ecolyte PS (used in food packaging) is a modified polystyrene containing carbonyl groups and is rapidly disintegrated in sunlight to a polymeric powder. Most vinyl polymers can be rendered sensitive in this way by the incorporation of about one percent carbonyl groups although subsequent decay is not assured. It has been estimated that the complete mineralization of Ecolyte PS may take between 10 and 100 years, although this may be unimportant because there is no evidence that the powdered photodegraded material has any harmful effect. Two other photodegradable plastics (Ecolyte PP—a modified polypropylene and Ecolyte EE, a modified polyethylene) are in fact completely mineralized by soil microflora after about a year. These polymers are used for food packaging in Canada and a photodegradable polythene is marketed in Sweden.

CONCLUSIONS

The development and production of synthetic polymers has, without doubt, contributed to improvements in the living standards of many people. Environmental problems are mainly concerned with those polymers which are used for tyres, disposable containers and wrapping materials. The use of plastics for protecting food has greatly reduced microbial contamination and deterioration, benefits which at present outweigh the disadvantages of having to dispose of the packaging after use. Nevertheless, the proportion of domestic and industrial waste accounted for by compressed plastics alone is steadily increasing and in most industrial nations is about three percent by volume (1.5% by weight) whilst in Japan it is approaching ten percent. Since these substances are, for the most part, only very slowly degraded they become widely distributed in the environment as litter. Litter is a social problem and can to some extent be controlled by a change of attitude coupled to efficient waste disposal systems. Nowadays a large proportion of synthetic polymer waste is used in "land fill" operations and this build-up of virtually indestructible material is regarded as undesirable by many.

At present synthetic polymer wastes do not pose any serious biological threat. However, since it seems clear that the amounts of packaging materials used will increase, it is important to take advantage of the developing technology in degradable plastics and fibres. Whether it will be possible to produce a plastic that is rapidly biodegraded remains to be seen but possibly such a substance would not show the excellent water-proofing properties of current packaging plastics. The answer probably lies in the further development of compounds which are photodegraded to products that are then rapidly and completely mineralized in the soil, either by further chemical reactions or by soil microorganisms.

Recommended Reading

Eggins, H. O. W., Mills, J., Holt, A. and Scott, G. (1971). Biodeterioration and biodegradation of synthetic polymers. *In* "Microbial Aspects of Pollution" (G. Sykes and F. A. Skinner, Eds.), p. 267. Academic Press, London.

Scott, G. (1973). Improving the environment: Chemistry and plastics waste. *Chemistry in Britain* 9, 276.

Staudinger, J. J. P. (1970). "Plastics, waste and litter." Society of Chemical Industry Monograph No. 35. London: Society of Chemical Industry.

Walsh, J. H. (1972). The plastics disposal problem. Biologist (Journal of the Institute of Biology) 19, 141.

"Degradability of Polymers and Plastics" (1973). The Plastics Institute, London.

7

Metals

INTRODUCTION

About one third of the Earth's ninety or so naturally occurring elements have been shown to be essential for the growth and development of microorganisms, plants and animals (Table 7.1). Some are both present and required in large quantities (carbon, hydrogen, oxygen, nitrogen, etc.) and are termed macronutrients. Others, the micronutrients, require sophisticated analytical procedures to detect them at all and may contribute less than 0.01 ppm to the dry weight of the living cell. In addition, we cannot be sure that certain organisms do not require the minutest quantity of an element hitherto regarded as non-essential.

Several of the elements listed in Table 7.1 become toxic if

TABLE 7.1

Essential nutrients of plants, animals and microorganisms

Macronutrients	Micronutrients
Hydrogen	Boron
Carbon	Fluorine
Nitrogen	Silicon
Oxygen	Vanadium
Sodium	Chromium
Magnesium	Manganese
Phosphorus	Iron
Sulphur	Cobalt
Chlorine	Copper
Potassium	Zinc
Calcium	Selenium
	Molybdenum
	Tin
	Iodine

available in excess of the organism's metabolic and physiological requirements. On the other hand, many plants and animals develop a tolerance to tissue levels of a particular metal far in excess of that required for "normal" growth. Notwithstanding, these resistant organisms may cause environmental problems by passing on the accumulation to more susceptible species.

The industrial nations have augmented the natural levels of available metals to such an extent that some are now present in the environment in quantities which may prove toxic to an whole range of animals and plants. Since the mid-1950's concern about the release of some elements, either as residues from industrial processes or in domestic waste and consumables, has been focused on the heavy metals, mercury and lead. Both of these are highly toxic elements with no known beneficial biological function. In addition, essential micronutrients (zinc, boron, selenium, copper) have attracted attention where they are found in abnormally high concentrations as have others such as arsenic, chromium, cadmium, beryllium, and nickel.

It is apparent that some forms of metals are more toxic than others. For example, organic mercury and lead compounds present an environmental hazard in excess of their equivalent inorganic forms and hexavalent chromium is more toxic than the trivalent state. Thus, generalizations concerning the toxicity of a particular metal have to be viewed very carefully. Even if we restrict the discharge of metals to those forms which are considered "safe" it is incorrect to assume that there will be no ecological recriminations, for many metals undergo biologically chemically and physically induced transformations which may convert non-toxic compounds into toxic ones.

The microbial oxidation and reduction of metals has been known for some time and is used commercially in the solubilization and leaching of certain elements. This is often an indirect process dependent, for example, upon the microbe producing an acid which acts on the insoluble metal salt converting it to a soluble form. On the other hand, many oxidation reactions yield substantial amounts of energy and the potential exists for exploitation by chemolithotrophs.

This chapter describes in detail the biochemistry, microbiology and toxicology of lead and mercury and briefly mentions other metallic pollutants.

ANALYTICAL METHODS

There are three major techniques currently used for the measurement of trace amounts of metals in the environment; colorimetry, atomic absorption spectroscopy and neutron activation analysis. There are also several other methods which are useful in certain instances; namely x-ray absorption and emission-mass spectrometry, polarography and gas chromatography. All methods vary considerably in their precision, sensitivity and reproducibility; this being particularly apparent in the case of mercury where there are considerable limitations on the detection of levels less than 0.1 ppm. As a result, marked differences in the residue values obtained for the same material using different methods (or even the same method in different laboratories) have been reported.

1. Special Precautions

Because of the minute quantities of metals being measured elaborate precautions are necessary in order to obtain reproducibility. Failure in this respect is one of the most likely causes of conflicting results and great care must be taken to avoid the introduction of extraneous metal ions during preparation of the sample and its subsequent analysis. The possibilities for error are numerous and include contamination from standard laboratory apparatus (e.g. zinc from rubber, copper and zinc from gas burners, arsenic, zinc and lead from glassware, chromium from acid-washed glassware, lead from porcelain and a variety of metals from filter papers). In addition, laboratory atmospheres may sometimes contain mercury vapour which can condense on apparatus being used for assays. Care must also be taken to trap dust particles which often have an high arsenic content. Even if all these sources of contamination are avoided inaccurate analyses may result from the adsorption or precipitation of the metal onto glass or filter papers.

2. Preparation of Samples

Sample preparation procedures depend both on the type of material involved and the subsequent analytical techniques. Plant and animal tissues are usually ground and homogenized followed by the destruction of any organic matter by ashing or wet oxidation. Ashing involves heating in a furnace at around 500°C and, although a simple method, suffers from the dual disadvantages that a proportion of some metals (particularly those which are volatile) is lost from the sample and that contamination from the porcelain dish and the furnace may occur. In addition, some metals may become adsorbed onto porcelain at high temperatures. The alternative method of removing the organic fraction, wet oxidation, involves heating at lower temperatures (about 300°C) with concentrated mineral acid mixtures (often sulphuric and nitric).

3. Assay Procedures

3.1. Colorimetric

Colorimetric methods are based on quantitative reactions which metal ions give with specific reagents (Table 7.2). Heavy metals are

TABLE 7.2
Colorimetric trace metal analysis

Reagent	Formula	Element assayed
Inorganic		
Persulphate	HSO_4^-	Mn, Cr, Os, Ru, I_2
Hydrochloric acid	HCl	Fe, Cu, Co, Pt
Hydrogen peroxide	H_2O_2	Mo, Ti, Cr
Organic		
Dithizone	$S{=}C(NH{-}NH{-}C_6H_5)(N{=}N{-}C_6H_5)$	Cd, Co, Cu, Pb, Hg, Zn
Diphenylcarbazide	$O{=}C(NH{-}NH{-}C_6H_5)_2$	Cr, Pb
8-Hydroxyquinoline	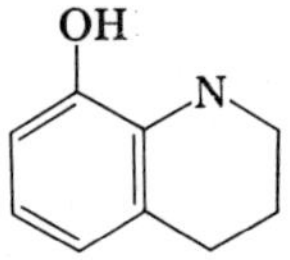 [structure: OH, N]	Al, Fe, Mg, Zn
Diethyldithiocarbamic acid	$(C_2H_5)_2N{-}C({=}S){-}SH$	Cu, Ni
Toluene-3,4-dithiol	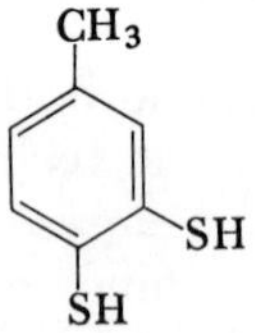 [structure: CH_3, SH, SH]	Sn, Mo
Thiocyanate	CNS^-	Fe, Co, Mo
Rubeanic acid	$HN{=}C(SH){-}C(SH){=}NH$	Co, Ni
Dimethylamino benzylidenerhodamine	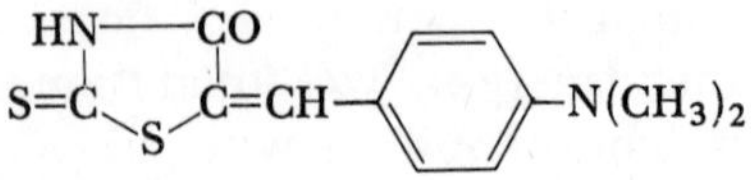 [structure: HN—CO, S=C, S, C=CH—C_6H_4—$N(CH_3)_2$]	Hg, Ag

particularly amenable to this type of analysis which for some is sensitive to 0.1 ppm (in biological samples). The main disadvantages of colorimetry in this sort of analysis are its lack of specificity and its sensitivity to impurities. It is usually necessary, therefore, to carry out elaborate separation and purification procedures (e.g. distillation, precipitation, co-precipitation, adsorption, extraction by immiscible solvents) before the final assay. This may lead to a pure sample but it is laborious and subject to extraction losses.

3.2. Atomic Absorption Spectrophotometry

After dry ashing or wet oxidation the sample is vaporized (in a flame) and the absorption of light at the appropriate wave length measured (Table 7.3). The absorbance values are then related to a

TABLE 7.3

Trace metal analysis by atomic absorption spectrophotometry

Element	λ (nm)	Detection limit (ppm)
As	194	0.3
Cd	229	0.002
Cr	360	0.01
Co	241	0.004
Cu	325	0.001
Fe	248	0.02
Pb	217	0.03
Hg	254	0.19
Ni	232	0.02
Sn	224	0.2
Zn	214	0.001

standard curve. The method is both accurate and versatile but there is considerable variability in its response to different metals. For example, it is extremely sensitive for zinc analysis (0.001 ppm) but relatively insensitive for mercury (0.19 ppm).

3.3. Neutron Activation Analysis

Neutron activation analysis is an highly sensitive method (0.0005 ppm) involving irradiating the sample in a chain reaction pile for two to four days during which it is bombarded with neutrons. As a result of this treatment most metals yield radioactive species. A standard, containing a known amount of the element in question, is exposed at the same time and the amount of metal in the sample determined from the ratio of the activities. It is often necessary to separate the particular element after irradiation as other elements in the sample may also become "activated".

3.4. Gas Chromatography

Gas chromatography is discussed in some detail in Chapter 2. It is a particularly valuable technique for the measurement of methyl mercury in biological material. The sample is hydrolysed with hydrochloric acid and the mercury isolated and purified as methyl mercuric chloride (CH_3HgCl) which can then be measured gas chromatographically using an electron capture detector.

3.5. Mass Spectrometry

Mass spectrometry is extremely sensitive for some elements (0.001 ppm) but the equipment required is expensive. A known amount of an isotope of the metal is added to the sample, the element chemically separated and the amount calculated from the ratio of the isotopes revealed by the mass spectrometer.

LEAD

1. Introduction

The world production of lead from basic ores at the beginning of the 1970's was almost four million tonnes, the largest contributors to this total being the U.S.A., Australia, the U.S.S.R. and Canada. Other countries, such as the U.K., whilst no longer mining lead, import large quantities for refining purposes. The subsequent smelting of the ore in a blast furnace, with limestone and iron, reduces it to metallic lead, carbon monoxide and iron sulphide.

Forty per cent of the world's refined lead is used in electrical batteries—either as components of alloys or as oxides. Cable sheathing accounts for another 25% whilst sheet and pipe lead find numerous uses in the building industry. Lead carbonate ($PbCO_3$) or white lead has been used extensively as a paint pigment in the past and is still a component of wood primers. Lead chromate is a yellow pigment in paint. Triplumbic tetroxide (Pb_3O_4) or red lead and calcium plumbate ($CaPbO_4$) are employed as iron and steel rust-inhibitors. Lead bisilicate is used in the glazing of ceramics, whilst tri-basic lead sulphate functions as a stabilizer of P.V.C. plastics. Tetraethyl and tetramethyl lead are added to petrol as "anti-knock" agents to improve the octane rating. These and other uses of lead are detailed in Table 7.4 from which it is apparent that the potential for environmental contamination is enormous. It is fortunate, therefore, that it is economically feasible to recycle large quantities of lead (especially from car batteries). Nevertheless, the high toxicity of lead and its build up in the environment have been the cause of much concern in recent years.

TABLE 7.4

Consumption of lead in the U.K.—1970

	Tonnes
Batteries	96 727
Cables	61 091
Sheet and pipe	54 418
Anti-knock Compounds	40 727
Alloys including solder	34 619
Shot	6 109
White lead	3 055
Miscellaneous metallic	22 400
Miscellaneous chemical	32 127

2. Chemistry

Lead is a bluish-grey, malleable element which exhibits a metallic lustre when its surface is clean. Exposure to air builds up a dull grey surface layer of hydroxide and carbonate, protecting the metal against further attack. Lead sulphide (PbS) or galena is the most important commercial source of lead whilst cerrusite ($PbCO_3$) and anglesite ($PbSO_4$) are mined in some areas. Lead occurs in divalent (plumbous) and trivalent (plumbic) states.

3. Distribution in the Environment

The natural dispersal of lead is somewhat restricted due to its insolubility. However, lead has been used and subsequently discarded by man in many forms for hundreds of years and must now be considered to have a global distribution.

3.1. Soil

Analyses of uncontaminated Antarctic soils show an average lead level of 10 ppm. This is regarded as the natural background figure for the element in surface soils although, in industrial and urban areas, levels are often greater than ten times this amount. Lead may arrive in the soil from deposition of aerosols and contaminated surface waters, as well as the limited mobilization of naturally occurring forms of the element. Soils nearest the source of output (beside busy roads or near smelters) will occasionally show concentrations greater than 1000 ppm in their surface horizons. The lead levels in sewage sludge may vary between 2000 and 8000 ppm and its use as a fertilizer may give rise to the subsequent contamination of agricultural soils (other heavy metals also concentrate in sewage sludge).

3.2. Water

Natural sources of lead contribute to aquatic contamination only to a small degree. The level is around 0.5 $\mu g\ L^{-1}$ in the surface layers of fresh water and less than 0.1 $\mu g\ L^{-1}$ in the open ocean (although, in the latter instance there is evidence that levels have risen significantly since the 1920's). Rivers in industrial regions often contain relatively large amounts of lead, notably at the source of effluent input (100 $\mu g\ L^{-1}$). Chronological ice horizons from Greenland indicate that lead concentrations have increased from 0.0005 $\mu g\ Kg^{-1}$ of ice in 800 B.C. to greater than 0.2 $\mu g\ Kg^{-1}$ in 1965.

3.3. Air

Atmospheric lead concentrations range from as low as 0.0002 $\mu g\ m^{-3}$ to as high as 71 $\mu g\ m^{-3}$ recorded during a peak traffic period on a Los Angeles freeway. Daily averages in cities are usually in excess of 3 $\mu g\ m^{-3}$. Wind, rain, water and air temperatures all influence the build up of pockets of abnormally high lead concentrations in the atmosphere.

4. Ecological Considerations

4.1. Effect of Lead on Microorganisms

Lead is normally quite toxic to microorganisms and much research has centred around this characteristic. The retardation of heterotrophic organic matter decay, following lead treatment, may be due to the competitive ability of the ions in binding to the $-SH$, $-NH_2$ and $-NH$ groups of enzymes, to the exclusion of such essential elements as manganese, iron and magnesium. Nevertheless, viable species of *Arthrobacter* and *Hyphomicrobium* have been isolated from the surface of lead ores, although there is no suggestion that there is an autotrophic relationship.

4.2. Effect of Lead on Macro-organisms

4.2.1. Plants. Lead has no known value, even as a micronutrient, to plants. Despite this, many plants are not only tolerant of high soil lead levels but tend to remove what ions there are and transport them through their root system and into the overground portions. This is clearly a genotypic characteristic as the same species transplanted from an uncontaminated soil will fail to grow. For example, grass in lead mining areas has been observed to contain 2000 ppm; wild oats (*Avena fatua*) near a smelter 500 ppm and pasture grass 9645 ppm. Vegetation near highways accumulates both a surface deposit of particulate lead and a systemic level derived from the soil. This effect decreases rapidly with distance from the road and is obviously dependent upon traffic density. Other species resistant to lead include ribwort *(Plantago lanceolata)*, creeping bent *(Agrostis stolonifera)* and red fescue (*Festuca rubra*).

The ability of plants to accumulate lead has been used to investigate changes in environmental levels over the years. Ideally one should conduct all the analyses at one time and with one method. Unfortunately comparisons are often made between figures arrived at many years ago, when analytical methods were poorly developed, and those achieved today with the most sophisticated equipment. It is not surprising, therefore, that many results appear contradictory. Two examples which are used today, for realistic comparisons, are tree rings and mosses. Specimens of mosses collected in Sweden between 1860 and 1875 show levels of 20 ppm. This concentration is seen to double in plants from the following twenty five years (1875-1900) and then remain fairly constant until 1950. In the last twenty years another dramatic increase has been observed and levels of between 80 and 90 ppm are recorded. It is suggested that the first rise coincided with the increased use of coal; the second with combustion of leaded petrol.

A change in environmental lead concentrations may also be inferred from measurements of tree ring deposits. Once such example describes an elm tree, growing some fifty metres from a street with light traffic, which showed the following trend: 1865-1879, 0.16 ppm; 1900-1912, 0.12 ppm; 1940-1947, 0.33 ppm; 1956-59, 0.74 ppm; 1960, 3.90 ppm. Remember that all these analyses were performed at the same time using the same techniques and thus if we consider absorbed lead to be immobilized in annular rings these results dramatically illustrate the changes in environmental lead levels over the past one hundred years.

4.2.2. Animals. Lead may be taken up by animals, either from food and water (10% retention) or from the air they breath (50% retention). For a number of years there has been concern about the possible health risks to the population due to the increasing contamination of the environment with lead from industrial processes, fossil fuel combustion, etc. However, the contradictions in the literature indicate that the harmful effects are poorly understood by some and considered negligible by others. Much of this confusion may be due to the complex toxicology of lead.

Inorganic lead (Pb^{2+}) is a general metabolic poison which may accumulate in a number of body tissues such as erythrocytes, liver and kidneys. It inhibits enzyme systems (delta-amino laevulinic acid (ALA) dehydrogenase) necessary for the formation in bone marrow of haem, the pigment which combines with protein to make haemoglobin. Lead also replaces calcium in bone, where it may remain for some time being re-mobilized at times of high calcium metabolism (illness, cortisone therapy, old age). Children appear, in some instances, more susceptible to lead poisoning—a factor recognized in the assessment of toxic levels in the blood which for adults is regarded as 0.8 $\mu g\ L^{-1}$ but for the children only 0.25 $\mu g\ L^{-1}$. Symptoms of lead poisoning range from mild (headache, fatigue, constipation, mild anaemia) to severe (nephritis, encephalopathy) and modern chelation therapy can help to reduce the effects.

Lead alkyls are even more poisonous than inorganic lead and induce a range of different responses in the body.

Traditionally, dangers of lead poisoning have come from occupational exposure—such as the inhalation of dust and fumes by people working in factories where lead is smelted or refined. In addition, the habit of children in chewing toys or furniture (pica) decorated with lead-containing paints has caused cases of poisoning. The now decreasing number of lead water pipes and joints has, in the past, contributed to high levels in drinking water in some areas and subsequent poisoning has resulted. Much of the present day concern about lead exposure arises from the presence of lead in the exhaust fumes of cars using petrol to which has been added lead alkyl "anti-knock" agents. Organic lead compounds— tetraethyl and tetramethyl lead—were first added to petrol in 1923 to increase octane numbers so that cars with high compression ratios could produce a high performance without abnormal combustion (knock). The amount of lead in petrol averages 2.6 g per U.S. gallon of which between 25 and 75% is exhausted as particulate material (less than 1 μm), soluble lead halides, oxyhalogenates and lead alkyls. These may affect plant and animal life directly and even affect weather conditions downwind of emission zones due to formation of so-called "ice-nuclei".

If lead is removed entirely from petrol (if that is considered desirable—and there are many who would disagree), many cars would not function satisfactorily without retarding the ignition and, in some cases, reducing the compression ratio. These two changes give rise to reduced performance, increased fuel consumption and increased emission of other pollutants—hydrocarbons and carbon monoxide. In the U.S.A. low lead and non-lead petrols were introduced in 1970. Present regulations are designed to achieve a 60-65% reduction in the use of lead alkyls by 1977 and encourage the production of automobile engines with lower compression ratios. The reduction in lead will also allow a more efficient use of catalytic devices to clean-up the exhaust, catalysts which were previously poisoned by lead.

MERCURY

1. Introduction

The mystical, medical, cosmetic and toxic properties of mercury have been appreciated by man for at least 2500 years. As early as 700 B.C. the same Spanish mines, which today supply a large proportion of the world's needs, were being worked by the Greeks. Aristotle (350 B.C.) writes about the use of mercury in religious ceremonies and the Arab physicians of the sixth century B.C. found medical functions for mercury compounds. Around the time of Christ the Greeks were using mercury for the treatment of skin diseases and its value in the treatment of syphilis was recognized in the sixteenth century.

The toxic qualities of mercury have been suggested as being responsible for the deaths of such notables as Napoleon, Ivan the Terrible and Charles II of England. More recent cases of mercury poisoning have occurred in Japan, Iraq, Pakistan and New Mexico with tragic results.

The unique properties of mercury led, in part, to its overwhelming appeal to Chinese, Arabian and European alchemists from the first century A.D. until their disassociation with the main stream of chemical thought in the 1600's. Notwithstanding, as late as 1750 there were still many who believed that mercury could be transmuted to gold.

Mercury is a natural component of the environment occurring as metallic mercury and mercuric sulphide (HgS). Additional contributions are made to this "background level" by a whole range of industries (Table 7.5). For example, the electrical industry uses

TABLE 7.5

Consumption of mercury in the U.S.A.–1969*

	Tonnes
Chlor-alkali plants	715
Electrical equipment	628
Paint	336
Instruments (thermometers, etc.)	178
Dental preparations	95
Agriculture	93
Laboratory use	57
Pharmaceuticals	24
Pulp and paper making	19

* From: Mercury in the Environment. Environ. Sci. Technol. 4, 890 (1970).

mercury in the production of batteries, street lamps, circuit breakers and relays with liquid contacts, all of which are discarded at the end of their useful life. Mercuric chloride catalysts are employed in the manufacture of vinyl chloride, urethane plastics and acetaldehyde. Phenyl mercury acetate is a fungicide in paints whilst a number of organic and inorganic mercury compounds are used in agriculture (Chapter 2). The largest contribution to environmental mercury levels is made by the chlor-alkali industry where a plant producing 100 tonnes of chlorine per day (electrolytically from sodium chloride using mercury as the cathode) may release between four and eight thousand kilograms of mercury per year in waste water. Other contributions are made by fossil fuels, paper and pulp industries, dental amalgams, antiseptics and floor waxes. In addition, the use of mercury as a fabric softener has given rise to the expression "mad as a hatter"–originally a reference to the toxic effects of mercury on workers in the hat making industry. World production (of which Spain, U.S.A., U.S.S.R., Italy and China are the largest contributors) is estimated at greater than 9000 tonnes per annum, of which some 50% may be "lost" to the environment.

2. Chemistry

Mercury is the only metal which is liquid at room temperature and it was not until the middle of the eighteenth century that it was accepted as a true metal. It is a silvery-white liquid which is slightly soluble in water, melts at –39°C and boils at 357°C. In between these temperatures its cubical coefficient of expansion is fairly uniform and this feature, combined with its inability to "wet" glass, makes it ideal for use in thermometers. Its low vapour pressure and high density makes it suitable for barometers.

Native mercury is occasionally found as small globules in rocks and as an alloy with gold and silver. Only one ore, however, is worth commercially exploiting: the red/black mercuric sulphide known as cinnabar, found in shallow mines in Southern Spain (Almaden), Italy (Tuscany, Trieste), U.S.S.R. (Ukraine), U.S.A. (California, Nevada), Yugoslavia and Mexico.

Mercury occurs in both monovalent (mercurous) and bivalent (mercuric) states and readily forms complex salts such as $K_2(Hg(CN)_4)$ and $K_2(Hg(SCN)_4)$. A range of organo-mercurials are found in microorganisms, plants and animals, alkoxy-, aryl-, ethyl-, methyl- and alkyl mercurials being the major groups. The biological transformations which give rise to these compounds are described later in the chapter.

3. Distribution in the Environment

Both elemental mercury and mercuric sulphide are extremely volatile. Thus it is not surprising that traces of mercury appear everywhere and that, with a little help from man, it cycles readily between soil, water, air, plants and animals.

3.1. Soil

The average concentration of mercury in the Earth's crust is estimated at 0.5 ppm, tending to be concentrated in sedimentary rather than igneous rocks. Reports of mercury levels in top soils are extremely variable ranging from 0.01 ppm to 2.0 ppm and may be much higher around areas of industrial and agricultural usage.

3.2. Water

Due to the dilution effects and adsorption to sediments, soluble mercury in sea water is rarely above 3×10^{-4} ppm. In coastal areas, subject to mercury containing industrial waste, levels may exceed 3×10^{-3} ppm. Rainwater analysis has revealed figures around 2×10^{-4} ppm whilst surveys of mercury levels in the Great Lakes have indicated concentrations as low as 5×10^{-5} ppm (Superior) and as high as 1×10^{-3} ppm (Ontario).

3.3. Air

Native mercury enters the atmosphere through the action of volcanoes and earthquakes and evaporation from water and soil surfaces. The background level of mercury in the atmosphere under "normal" conditions has been variously estimated at less than 1 mg m^{-3}. However, localized natural deposits may increase this level as much as twenty fold. This characteristic has been used by prospectors to detect not only mercury ores but gold-bearing rocks in areas where a correlation between these two precious metals had previously been observed.

Man's industrial contributions dramatically augment atmospheric levels, especially in enclosed areas. In electrical and chlor-alkali factories the concentration of mercury in air may be in excess of 0.1 mg m^{-3} and even higher closer to the source of emission. For example, one measurement, in a workshop repairing direct-current meters, cites a mercury level of 1.6 mg m^{-3} directly over the work benches. This is some thirty times in excess of the maximum legal limit. Dental surgeries and rooms recently decorated with mercury-containing paints also show levels well in excess of the background.

4. Ecological Considerations

4.1. Transformations of Mercury

It is apparent that, whilst inorganic forms of mercury are extremely toxic to plants and animals, organic complexes may prove to be the long term hazard to our environment.

Mercury and its inorganic and organic compounds undergo a variety of chemical and biological transformations. For example, it is now well known that methyl- and dimethyl-mercury are formed in aquatic environments due to the activities of free-living bacteria and fungi. This methylation process has also been observed to be mediated by enzyme systems in rotting fish and lake sediments in addition to that caused anaerobically by some *Clostridium* species (especially *C. cochlearium*) and aerobically by the methanogenic fungus *Neurospora crassa* and members of the genus *Pseudomonas*. It is suggested that this is a chemical process in which methylcobalamin is involved [I] .

$$Hg^{2+} \longrightarrow CH_3Hg^{+} \qquad [I]$$

Microbes (especially bacteria) have also been implicated in (a) the reduction of phenyl-, ethyl- and methyl mercury to form elemental mercury (sometimes volatilized) and benzene, ethane and methane respectively, (b) the aerobic conversion of phenyl mercuric acetate (PMA) to elemental mercury and diphenyl mercury and (c) the reduction of Hg^{2+} ion to elemental mercury (*Pseudomonas*, *Enteric* bacteria and *Staphyloccoccus aureus*).

Non-biological transformations of mercury include the conversion of methyl mercury, in alkaline conditions, to the more volatile dimethyl mercury [II]. In acid environments methyl mercury remains in the water.

$$CH_3Hg \longrightarrow (CH_3)_2Hg \qquad [II]$$

Mercury, present as phenyl mercury, alkoxy alkyl mercury and alkyl mercury, may be broken down to inorganic mercury. Under anaerobic conditions mercuric ions may combine with biologically produced hydrogen sulphide (especially in eutrophic lakes) to form

$$H_2S + Hg^{2+} \longrightarrow HgS \qquad [III]$$

poorly soluble mercuric sulphide[III]. With subsequent aeration mercuric sulphide can be converted to a soluble sulphate which, may then, be methylated biologically [IV].

$$HgS + 2O_2 \longrightarrow HgSO_4 \longrightarrow CH_3Hg^+ \qquad [IV]$$

It is worth recording that conversions of many other mercury salts (e.g. mercuric nitrate, mercuric chloride, phenylmercuric acetate) have not been described.

Some of the transformations of mercury are outlined in Fig. 7.1.

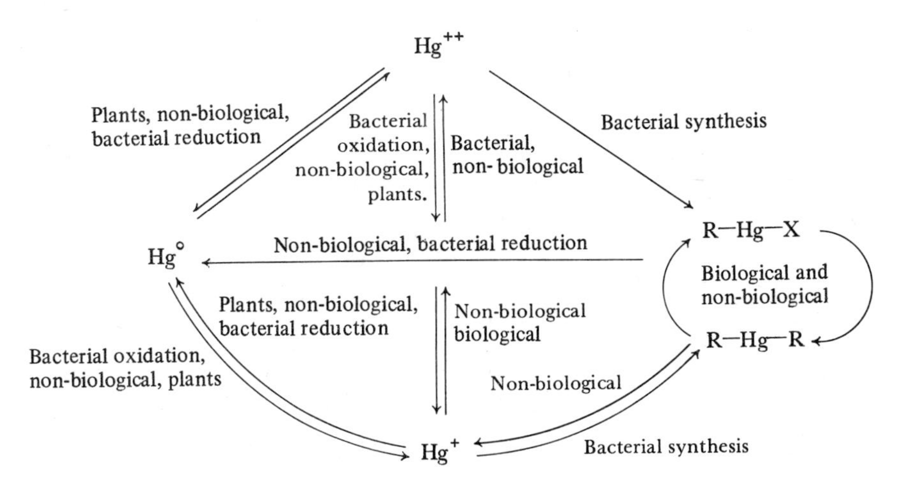

R,R′ = alkyl, aryl, mercapto, methyl, ethyl, etc. X = monovalent anion(acetate)

Fig. 7.1. Transformations of mercury.

4.2 Effect of Mercury on Microorganisms

The effect of mercury on the total number of soil microbes, their enzyme activities and their rates of carbon and nitrogen mineralization have been investigated. The organo-mercurials are often seen to be more inhibitory than their inorganic counterparts. For example, it takes some twenty times as much mercuric chloride to decimate the growth of microbes, grown on amended soil-extract agar plates, as it does using methyl mercury chloride. PMA and mercuric chloride, at concentrations of 100 ppm, inhibit dehydrogenase activity. Ammonification and, more dramatically, nitrification are retarded by 100 ppm of either mercuric chloride or PMA, although the organo-mercury appears the more potent and soil type is important in determining the range of response. Both marine and fresh water phytoplankton (especially diatoms) are sensitive to organomercurial fungicides and as little as 0.001 ppm may reduce photosynthesis. Many of the procaryotic algae accumulate cellular concentrations many times in excess of the mercury levels found in sea water.

When one considers that mercury compounds, used as pesticides, enter the soil in amounts of between 0.0005 and 1 ppm (1-200 g ha^{-1}) it is unlikely that the microbial changes induced in the experiments described here will occur under "normal" conditions.

4.3 Effect of Mercury on Macro-organisms

Traces of mercury occur in all plants and animals so far examined. Plants may absorb and concentrate mercury from soil, even to the extent that droplets of the metal have been found in the capsules of chick-weed *(Holosteum umbellatum)*. This build up of toxic metal may cause mitotic disturbances in the plant cell which may prove to be lethal. It is fortunate, therefore, that agricultural crops, grown on land sprayed with mercury-containing pesticides, appear to absorb only small levels and the resulting tissue concentrations are generally less than 0.1 ppm. Of course, translocation within plants is very much dependent upon solubility and it is the highly toxic methyl mercury which is most soluble.

Animals, which are involved in long food chains, tend to accumulate a level of mercury far in excess of that found in their immediate environment. Insectivorous birds, which feed on lacewings, which feed on aphids which, in turn, suck mercury-contaminated sap from plants may build up levels many thousand times that found in the pesticide treated soil. In fresh-water environments pike accumulate up to three thousand times as much mercury as in the water, whilst swordfish and tuna also have this unfortunate ability. Predatory birds, such as owls, eagles and falcons, are sometimes found to have greater than 100 ppm of mercury in certain organs (especially the liver and kidneys). This mercury may have come from natural sources but it is more frequently from seed dressings, industrial water pollutants, etc. The rapid rise in mercury levels in the feathers and eggs of seed-eating birds dramatically parallels the introduction of alkyl-mercury seed dressings in the mid 1940's. Previous use of inorganic mercury had no such effect. In 1966, when the alkylmercurials were replaced in some countries (Sweden, U.S.A. and Canada) by alkoxy-alkyl compounds, there was a rapid decline in the mercury levels of monitored birds.

The effects of mercury poisoning in man and its mobility through food chains are dramatically illustrated by what is now known as the "Minamata Incident". Minamata is a small industrial town on the west coast of Japan's southern most island, Kyushu. In an eight year "epidemic" (1953-61) well in excess of one hundred people suffered methyl mercury poisoning, for more than forty of whom it proved fatal. Such are the varied and complex symptoms of the disease that it is probable that many others experienced some degree of poisoning.

Initial investigations considered the possibility of bacterial or virus infections and a range of heavy metals (thallium, lead, manganese) as being the causal agents of "Minamata disease". The human deaths, in and around Minamata, coincided with large scale fish mortalities and, as a relationship between diet and illness appeared possible, a temporary ban was placed on fishing. Eventually mercury was identified as the cause and high levels of methyl mercury were found in bay mud (2100 ppm), fish and shellfish. The symptoms of the disease were then successfully recreated in cats, mice and rats fed on food containing organic mercury.

The over-riding question now was where had the mercury come from? To understand the difficulties behind forming, what appears today as an obvious association between a local chemical factory (Chisso) and the mercury poisoning, one has to realize that the transformation of inorganic mercury to organic mercury was unknown at that time. Therefore, although the Chisso plant used mercuric sulphate catalysts in acetaldehyde production, only the comparatively nontoxic inorganic mercury was released in the industrial effluent.

It was five years after the outbreak of mercury poisoning that organic mercury (as methylmercury chloride) was isolated from the factory waste water and the transformation discovered. Legislation was introduced to ensure the clean-up of the mercury-containing effluent but methods used were only partially effective and it wasn't until as recently as 1968 that the dumping of mercury waste was finally halted. The legal aspects of the case, particularly compensation to the crippled and the relatives of the dead, were only resolved in 1973—some twenty years after the first cases appeared and when the total fatalities had reached sixty-five.

Fishing in Minamata Bay is heavily restricted nowadays yet new patients continue to register (400 to date) as the long-term accumulation of mercury becomes evident. It is estimated that some 400 tonnes of mercury are on the bottom of the bay closest to the original source of effluent inflow. Much of this may be bound and unavailable but its transformation into the deadly methylmercury may be a continuing process.

CADMIUM AND ZINC

1. Introduction

Cadmium and zinc, belong to subgroup II of the periodic table and, as a consequence, share a number of physical and chemical characteristics. In addition they are usually mined and refined together for cadmium is essentially an impurity (in the ratio 1 : 200, Cd : Zn) in zinc ores such as zinc blend (ZnS). Only rarely does cadmium appear as an ore "in its own right"—as cadmium sulphide or greenockite (CdS).

Zinc is produced by either smelting the ore or by a combination of leaching and electrolytic reduction. Cadmium is more volatile than zinc and is concentrated in the first fraction of the distillate during smelting or it may be precipitated, along with the more noble metals, in the electrolytic process.

2. Cadmium

Cadmium is a natural component of soil (0.5 ppm) and water (0.4 $\mu g\ L^{-1}$) where it exhibits a low solubility. In areas where there are significant zinc deposits and in active mines, water levels of cadmium may be in excess of 3 $\mu g\ L^{-1}$. In addition its world-wide industrial use is expanding—6000 tonnes in 1950, 18 000 in 1970. Cadmium is used as an antifriction agent (bearings), a rust proofer (especially in paints and varnishes), an alloy (with lead, copper) and has many other functions.

The ability of plants to absorb and concentrate cadmium means that it is a real danger to herbivores and a potential danger to carnivores. The daily intake in food by members of industrial societies is in the region of 200-400 μg. Fortunately retention is very low (1-2%) and accumulation is necessarily very slow. However, to this must be added atmospheric cadmium, which by all accounts is retained in larger quantities by the body. In industrial environments (smelters) the maximum permissible level is 100 μg m^{-3}.

TABLE 7.6

Pathological effects of cadmium

Pathology	Daily intake (in μg) for 1–3 years
Hypertension	175
Anaemia	530
Retardation of growth	1300
Abnormalities of pancreas and spleen	1300
Heart abnormalities	2125
Liver and kidney damage	5250

As far back as 1957 Scandinavian toxicologists had recognized the potential dangers of cadmium accumulation in the kidneys. Experiments during the 1960's with rats showed that continuous exposures to cadmium in drinking water induced a permanent hypertension, heart enlargement and a 20% reduction in life-span. Other effects, related by some to environmental cadmium levels, include cirrhoses of the liver, lung damage and destruction of testicular function (Table 7.6).

In 1955 there was an outbreak of cadmium poisoning in Northern Japan associated with the high metal content of rice and soya bean (0.37-3.36 ppm dry wt.). Called the Itai-Itai-Byo disease the symptoms were typical of extreme lumbago—painful lack of mobility followed, in some patients, by skeletal collapse. In these severe cases cadmium was apparently increasing bone porosity and inhibiting bone repair mechanisms. Between the years 1962 and 1968 there were over 200 reported instances (mostly female) of this type of cadmium poisoning in Japan.

In the light of this knowledge, the use of sewage sludge containing 100-500 ppm cadmium, as an agricultural fertilizer, has generated considerable controversy.

3. Zinc

Unlike the previously mentioned metals, zinc is an essential micronutrient for microorganisms, plants and animals. In high concentrations zinc is normally toxic to living systems, although there is evidence that some plants accumulate as much as 15% of their dry weight as zinc and are unaffected.

Levels of zinc in the "natural" environment are of the order of 10 $\mu g\ L^{-1}$ (river water) with an average of 40 ppm overall in the Earth's crust. Water flowing from areas of significant zinc deposits may contain greater than 250 $\mu g\ L^{-1}$.

Zinc plays an essential role in physiological and metabolic processes of many microorganisms. For example, it is needed in (a) the syntheses of many enzymes and their amino acid components—e.g. in *Aspergillus niger*—phenylalanine, tryptophan, tyrosine, phosphofructokinase and nucleic acids, (b) the production of antibiotics—penicillin (*Penicillium*), subtilin (*Bacillus subtilis*), bacitracin (*Bacillus licheniformis*), (c) cytochrome syntheses (*Ustilago sphaerogens*), (d) the control of decreased organic acid synthesis coupled with increase in cell yield known as the zinc shunt (*Aspergillus, Rhizopus, Penicillium*), (e) nitrogen fixation (*Azotobacter*), (f) mutant production and sporulation (many fungi) and (g) several enzymic reactions—phosphatase, polypeptidase (*Clostridium, Propionibacterium*), oxalacetate decarboxylase (*Azotobacter vinelandii*).

Since zinc does not have several common valency states it does not offer oxidative microorganisms an easily available energy source. Zinc sulphide (the commonest ore) however, has a sulphide moiety which is oxidized under acid conditions to sulphate, (by *Thiobacillus ferrooxidans*, for example). The resulting soluble zinc sulphate is then leached from the source or deposit. This form of bacterial oxidation is probably initiated through the generation of ferric sulphate which promotes the dissolution of sulphides (Fig. 7.2) although direct microbial oxidation of zinc sulphide should not be discounted.

$$ZnS + 2Fe_2(SO_4)_3 + 2H_2O + O_2 \rightarrow ZnSO_4 + 4FeSO_4 + 2H_2SO_4$$

Fig. 7.2. Bacterial oxidation of zinc sulphide.

NICKEL

The most important nickel ores-pentalandite (NiS . 2FeS) and garnie rite (Ni . Mg) $SiO_3 . nH_2O$—are refined either by smelting (the Mond process) or electrolytically. The toxicity hazards arising from smelting are twofold as the nickel carbonyl ($Ni(CO)_4$) is decomposed to produce both carbon monoxide and nickel. Nickel, itself, may cause lung complaints amongst workers who are continuously exposed.

The background level of nickel in the Earth's crust is around 200 ppm whilst clean streams are found to contain in the region of 50 ppm. Increases above this level may be caused by leaching from natural deposits, deposition of particulare nickel (70 000 tonnes/year reach the atmosphere from fossil fuels) or industrial processes—purification and alloy production. Nickel finds numerous uses as an alloy in combination with iron and carbon (nickel-steel, platenite), chromium and iron (nichrome), copper (nickel coins, constantan), copper and zinc (German silver) and others such as manganese, aluminium and molybdenum.

Some plants seem to accumulate nickel although none have an obligate requirement. Nickel has been observed to enhance sporulation of *Bacillus coagulans* and may replace calcium (presumably in calcium dipicolinate) in *Bacillus megaterium* spores. Nickel also stimulates aryl sulphatase activity in some *Pseudomonas* species. Microbial transformations of nickel may include the formation of nickel sulphide from nickel carbonate and nickel hydroxide after reaction with microbiologically produced hydrogen sulphide.

CHROMIUM

Chromium was first discovered by a French chemist in 1797 and is today mined as chrome iron ore ($FeO \cdot Cr_2O_3$) in Turkey, Africa, U.S.S.R. the Philippines and U.S.A. Background levels are in the region of 400 ppm (earth's crust), 0.04 ppm (river water) and 0.00005 ppm (sea water).

The variety of valency states (Cr^{2+}, Cr^{3+}, Cr^{5+} and Cr^{6+}) should make it suitable for microbial oxidation but no evidence exists for this. Non-biological transformations, mediated by heat and organic matter, do occur. Chromium is toxic in high concentration to both plants and animals. In continuously exposed humans it may lead to nasal perforations, bronchiogenic carcinomas and accumulate to levels far in excess of those found in the immediate habitat. Some algae may exhibit concentration factors as high as four thousand.

Other metals of commercial importance on which attention has been focused in recent years include copper, antimony, selenium, arsenic and beryllium. All of these are toxic to plants and animals at high levels but we know comparatively little about their biochemistry. As to the microbiology of these elements the leaching of copper minerals has been well documented, selenium has been suggested as an energy source for some autotrophs whilst a range of microbial genera are capable of methylating arsenic and oxidising arsenite to arsenate.

Recommended Reading

Bryce-Smith, D. (1971). Lead pollution—a growing hazard to public health. *Chemistry in Britain* 7, 54.

Bryce-Smith, D. (1973). Pollution of the soil by heavy metals. *Journal of the Royal Society of Arts*, February, p. 120.

Chow, T. J. (1973). Our daily lead. *Chemistry in Britain* 9, 258.

Goldwater, L. J. (1971). Mercury in the environment, *Scientific American* 224, 15.

Page, A. L. and Bingham, F. T. (1973). Cadmium residues in the environment. *Residue Reviews* 48, 1.

Saha, J. G. (1972). Significance of mercury in the environment. *Residue Reviews* 42, 103.

Smart, N. A. (1968). Use and residues of mercury compounds in agriculture. *Residue Reviews* 23, 1.

Wood, J. M. (1974). Biological cycles for toxic elements in the environment. *Science* 183, 1049.

Zajic, J. E. (1969). "Microbial Biochemistry", Chap. 8. Academic Press, New York.

8

Miscellaneous Pollutants

AIR POLLUTION

1. Introduction

Air pollutants can be conveniently divided into those arising from natural sources (volcanoes, organic matter decay, etc.) and those produced by man and his industrial society. In general, the ones with a natural origin make by far the greatest contribution to global atmospheric pollution.

Man-made atmospheric pollution has, of course, existed since his discovery of fire and attempts to limit his contribution have been made periodically ever since. Nowadays the major sources of air pollution include transport, industry, power plants, heating and refuse disposal together with minor (yet locally significant) outputs from plant and microbial allergens, spray aerosols, naturally occurring hydrogen sulphide and methane, and so forth.

The contribution of modern man to the composition of the atmosphere may be assessed by a comparison of Tables 8.1 and 8.2.

2. Types of Pollutant

2.1. Sulphur

World-wide sulphur emissions total 19.5×10^{10} kg per year and are mainly sulphur dioxide, hydrogen sulphide and sulphate (Table 8.3). It can be seen that natural sulphur (bacteria and sea spray) exceeds the input of man

2.2. Nitrogen

The major natural source of nitrogen gases is bacterial activity which contributes something like 98% of the total (Table 8.4). Nitric oxide

TABLE 8.1

Composition of clean, dry air*

Component	% by volume
Nitrogen	78.090
Oxygen	20.940
Argon	0.930
Carbon Dioxide	0.032
Neon	1.8×10^{-3}
Helium	5.2×10^{-4}
Methane	1.5×10^{-4}
Hydrogen	5×10^{-5}
Carbon Monoxide	1×10^{-5}
Ammonia	1×10^{-6}
Ozone	2×10^{-6}
Nitrogen dioxide	1×10^{-7}
Sulphur dioxide	2×10^{-8}

* 1969 Report Am. Chem. Soc. Washington D.C. "Cleaning our Environment".

TABLE 8.2

Pollutant composition of Chicago air*

Component	% by volume
Sulphur Dioxide	1.35×10^{-5}
Nitrogen Dioxide	4.2×10^{-6}
Carbon Monoxide	7.6×10^{-4}
Total Hydrocarbon	3.2×10^{-4}

* from Lynn and McMullen (1966). Air pollution in six major U.S. cities. JAPCA 16, 186-190.

TABLE 8.3

World-wide sulphur emissions*

Compound	Source	Kg yr^{-1}
SO_2	Coal	4.7×10^{10}
SO_2	Petroleum	1.3×10^{10}
SO_2	Smelting	7.0×10^{9}
H_2S	Bacteria	8.8×10^{10}
SO_4	Sea Spray	4.0×10^{10}

* Robinson, E. and Robbins, R. C. (1970). Gaseous sulphur pollutants from urban and natural sources. J. APCA 20, 233.

TABLE 8.4

World-wide nitrogen emissions*

Compound	Source	Kg yr^{-1}
NH_3	Bacteria	8.6×10^{11}
N_2O	Bacteria	3.4×10^{11}
NO	Bacteria	2.1×10^{11}
NO_2	Coal	7.4×10^9
NO_2	Space Heating	4.5×10^9
NO_2	Motor Vehicles	2.0×10^9
NO_2	Other Combustion	0.7×10^9
NH_3	Combustion	3.1×10^9

* Robinson, E. and Robbins, R. C. (1970). Gaseous nitrogen compound pollutants from urban and natural sources. J. APCA 20, 303.

is a relatively harmless gas except for its indirect effect as an oxidant. Oxidation to the pungent nitrogen dioxide, however, converts it into a serious health hazard.

2.3. Oxides of Carbon

Carbon monoxide, mostly from automobile exhausts, enters the Earth's atmosphere in quantities of 1.8×10^{11} kg yr^{-1}. It is a stable and highly toxic gas and, as such, is a dangerous pollutant in areas of high traffic density where levels may reach 70 $\mu g\ m^{-3}$.

It is unlikely that carbon dioxide has a direct effect on animal life but its more subtle relationships with Man may centre around its influence on climate (page 217).

2.4. Hydrocarbons

Hydrocarbons originate from natural organic matter decomposition (methane, terpenes) and the incomplete combustion of fossil fuels. Although man-made hydrocarbon emissions only represent a small proportion of the total (about 15%) they are usually concentrated in areas of high population density.

2.5. Particulate Matter

The particulate components of the atmosphere, mostly carried in the trophosphere (0-10 km), are those fragments varying in size from 0.1 μm to greater than 100 μm. Particles larger than 10 μm usually settle out quite rapidly due to deposition by gravity, rain and snow. In heavy industrial areas as much as 1000 tonnes of dust falls on each square mile during a year. Particles less than 1 μm in size are termed aerosols and may be transported vast distances especially if injected into the stratosphere (approximately 10-25 km). Particulate matter may arise from erosion, meteorites, volcanic dust, mechanical and industrial sources, combustion ash, motor vehicle exhaust, etc. It is estimated that over 90% results from natural sources.

2.6. Allergens

Allergic reactions, due to the over-production of histamine and related substances are sometimes described as the most important aspects of atmospheric pollution in relation to ill-health. Plants, animals and microorganisms generate a vast quantity of material that has the potential to trigger an allergic response. Hypersensitivity to ragweed pollen is widespread, a common cause of hay-fever, and responsible for the loss of countless man-hours of work.

2.7. Radioactivity

Both natural and man-made radio-active pollutants occur in the atmosphere. Natural sources include a range of radioactive elements produced by cosmic radiation–such as tritium, potassium-40, carbon-14 and beryllium-7. Three radioactive gases produced in atomic reactors are of concern, tritium, iodine-131 and more especially krypton-85.

2.8. Miscellaneous

Other atmospheric pollutants must include the unpleasant smells emanating from farm yards (e.g. ammonia), chemical factories, tanneries and sewage works together with bodily odour and cigarette smoke in enclosed areas. It is perhaps fortunate that our sense of smell is not as acute as that of many other animals.

3. Transformations of Gaseous Pollutants

3.1. Acid Rainwater

Oxides of sulphur, nitrogen and carbon may undergo chemical transformations to produce a variety of acids. This may prove to be a major environmental hazard especially in areas where "acid rainwater" has been shown to have a pH as low as two.

Sulphur dioxide may react with atmospheric water to produce sulphuric acid, or may first be oxidized to sulphur trioxide which, by virtue of its hygroscopic nature, combines with water. The commercial production of sulphuric acid (the lead chamber process) mimics the atmospheric interaction of water, sulphur dioxide and nitrogen dioxide whilst nitrogen dioxide, itself, reacts with water to form both nitric and nitrous acids. The latter is unstable and is decomposed to the former. Carbon dioxide dissolves in water to produce carbonic acid. The chemistry of these reactions is summarized in Table 8.5.

Table 8.5

Acid production in the atmosphere

Reactants		Products
$2SO_2 + 2H_2O + O_2$	$\longrightarrow$	$2H_2SO_4$
$\left\{ 2SO_2 + O_2 \right.$	$\longrightarrow$	$\left. 2SO_3 \right\}$
$\left\{ SO_3 + H_2O \right.$	$\longrightarrow$	$\left. H_2SO_4 \right\}$
$SO_2 + NO_2 + H_2O$	$\longrightarrow$	$H_2SO_4 + NO$
$2NO_2 + H_2O$	$\longrightarrow$	$HNO_2 + HNO_3$
$3HNO_2$	$\longrightarrow$	$HNO_3 + 2NO + H_2O$
$CO_2 + H_2O$	$\longrightarrow$	H_2CO_3

3.2. Photochemical Smog

Photochemical smog, of the Los Angeles–Chicago–Tokyo type, differs greatly from the more traditional London "pea soup" smog. London smog is a combination of fossil fuel pollutants and heavy natural fog unaffected by sunlight. By contrast, the components of photochemical smog are secondary pollutants formed by the catalytic activity of solar energy.

Ultra-violet light stimulates the dissociation of nitrogen dioxide to form two oxidants–nitric acid and dissociated oxygen[I]. In turn, atomic oxygen reacts with atmospheric oxygen to produce a third important catalyst–ozone[II].

$$NO_2 \rightarrow NO + O \qquad [I]$$

$$O + O_2 \rightarrow O_3 \qquad [II]$$

Hydrocarbons also react with sunlight to dissociate into chemically active free radicals which may then complex with the oxidants to form a wide range of secondary photochemical pollutants. These include aldehydes, ketones, peroxyacetyl nitrate or PAN ($CH_3(CO)OO . NO_2$) and peroxybenzoyl nitrate or PBzN ($C_6H_5(CO)OO . NO_2$).

The environmental dangers, due to the products of these reactions, are complexed by stagnant air and low-lying inversion layers and are clearly correlated to the concentration of motor vehicles.

4. Ecological Considerations

4.1. Effect of Air Pollutants on Plants and Microorganisms

The annual financial burden caused by air pollutants, in relation to loss of plant productivity, has been estimated to be as high as £50 million in the state of California and £2.5 million in industrial East Lancashire.

Particulate matter may reduce crop yield indirectly by restricting incident light, or directly by deposition on leaf and stem surfaces. In the latter instance stomata may be come blocked and transpiration significantly reduced.

Gaseous atmospheric components, such as sulphur dioxide, reduce the growth of a great many plants when at concentrations of around 60 $\mu g\ m^{-3}$, a level common in cities and surrounding areas during the winter months. Amongst the higher plants some (oak, larch, plane, maple, elder) are tolerant of air pollutants whilst others (pine, spruce, fir, bilberry) are not. Sulphur dioxide has also been reported as responsible for a reduction in the occurrence of two fungal pathogens of roses—the ascomycetes, *Diplocarpon rosae* causing back-spot and *Sphaerotheca pannosa* the rose-mildew fungus. It is possible that many other inhibitions of this nature occur. The microbial inter-relationship of an alga and a fungus, the lichen, is often disrupted by atmospheric pollution. Many investigations have illustrated a decline in lichen species with decreasing distance from urban centres (Fig. 8.1). It appears that epiphytes are most sensitive and terrestrial lichen forms are more resistant. In heavily polluted areas only a few crustose lichens survive, but, as pollution decreases fructicose forms appear until eventually foliose lichens grow where pollution levels are low.

Acid rainwater may "sour" a soil and severely restrict plant growth.

4.2. Effect of Air Pollutants on Human Health

The very real dangers of atmospheric pollution to human health are tragically illustrated by the five day London smog of December 1952, during which the death toll was some 4000 in excess of that statistically expected. Clearly, for many individuals already suffering from respiratory diseases, this event was the last straw.

The effects of individual gaseous pollutants (especially ozone, sulphur dioxide and nitrogen dioxide) include headaches, eye irritation and respiratory interference. Carbon monoxide has attracted much attention due to its high toxicity (ten times that of sulphur dioxide) arising from its ability to combine with the respiratory pigment of the blood, at the expense of oxygen, to produce carboxy-haemoglobin. Acute exposure to carbon monoxide rapidly produces dizziness, nausea, muscular weakness, collapse and ultimately death. Hydrocarbons as well as contributing to photochemical smog, may be directly toxic especially polycyclics such as benzpyrene, fluoranthrene, chrysene, coronene and anthanthrene. Particulate matter produces a variety of pneumoconioses (anthracosis, asbestosis, silicosis, farmer's lung etc.). The effects of toxic metal particulates are described elsewhere (Chapter 7). People living in areas of high atmospheric pollution show a predisposition to a variety of bacterial and viral infections.

4.3. Effect of Air Pollutants on Climate

The relationship of air pollutants to weather is a much studied and complex phenomenon. The formation and persistence of fog depends, to a large extent, on suspended particles onto which water can condense (condensation nuclei). Lead particles and iodine may induce rainfall and are, in fact, used by man for "cloud seeding". Pollutants may also effect the global heat balance either by blocking out radiation (particulates) or, conversely, by reducing reflection (carbon dioxide)—the so-called "greenhouse effect".

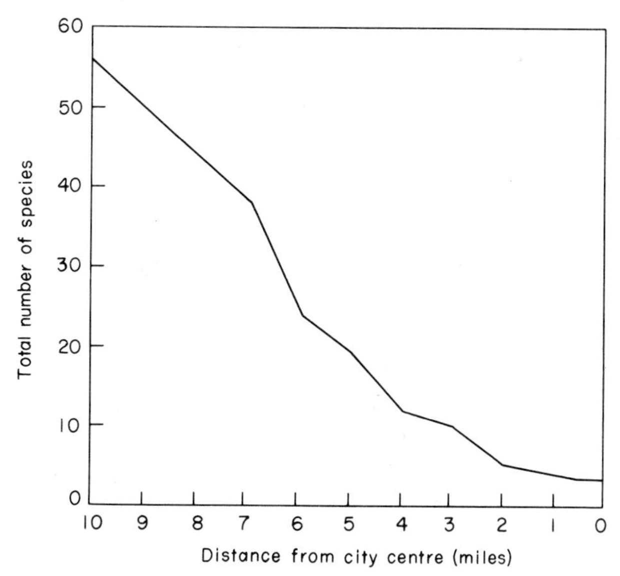

Data for Newcastle-upon-Tyne, England. (From Gilbert, O. L., (1965). *In* "Ecology and the Industrial Society". Ed. by G. P. Goodman, R. W. Edwards and J. M. Lambert. Blackwell Scientific Publications, Oxford).

Fig. 8.1. Decline of lichen species with increased proximity to the city centre.

ACID MINE WATER AND MICROBIAL CORROSION

Two important environmental hazards are observed during the biological cycling of sulphur—the production of acid drainage water and microbiological corrosion.

1. The Sulphur Cycle

Sulphur exists in the environment as mineral sulphates and sulphides, in organic matter, as fertilizers and soil pH treatments, as gaseous sulphur dioxide and as sulphuric acid in rainwater. As a component of plant, animal and microbial organic matter, sulphur (as sulphydryl, R—SH groups) provides a linkage between the polypeptide chains in the protein molecule helping to maintain its three-dimensional structure. Sulphur-containing organic compounds include the amino acids cystine, cysteine and methionine; B vitamins such as thiamine, biotin and thioctic acid and a variety of excretory products such as taurine, thiosulphate and thiocyanate.

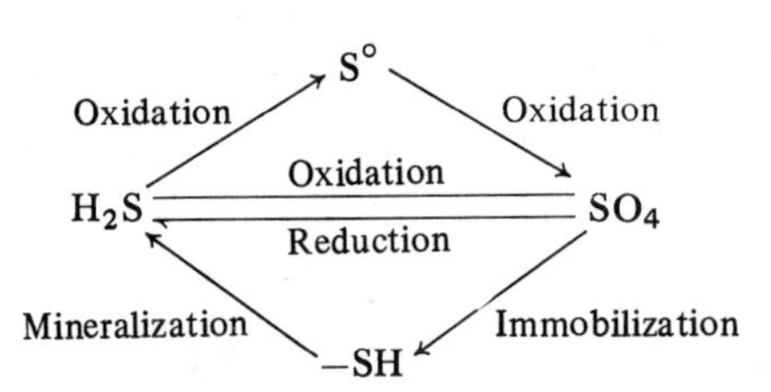

Fig. 8.2. The sulphur cycle.

The sulphur cycle (Fig. 8.2) may be conveniently discussed under the sub-headings mineralization, immobilization, oxidation and reduction. The biochemistry of these transformations is, in parts, highly speculative.

1.1. Mineralization

A wide variety of heterotrophic microorganisms are capable of decomposing sulphur-containing organic residues. A portion of this sulphur is used in cell synthesis whilst, in many instances, the remainder is released as hydrogen sulphide. Hydrogen sulphide is unstable in aerobic environments and is rapidly oxidized by chemical and biological mechanisms. Protein hydrolysis, with reference to the nitrogen cycle, has been described previously (Chapter 3). If a sulphur amino acid is released during proteolysis it may undergo microbial desulphydration. For example, cysteine is degraded by the enzyme cysteine desulphydrase (Fig. 8.3). Microorganisms reported as performing this hydrolysis include *Proteus vulgaris, Escherichia coli* and *Bacillus subtilis.*

$$HS{-}CH_2{-}CH_2{-}\underset{}{\overset{NH_2}{\overset{|}{C}}}H{-}COOH + H_2O \longrightarrow H_2S + NH_3 + CH_3\overset{O}{\overset{\|}{C}}{-}COOH$$

cysteine *pyruvate*

Fig. 8.3. Degradation of cysteine.

On other occasions cysteine-sulphur may be oxidized all the way to sulphate without involving sulphide by a combination of chemical and biological mechanisms.

Methionine, by contrast, may be oxidized to volatile intermediates such as methyl mercaptan (Fig. 8.4).

$$CH_3{-}S{-}CH_2{-}\overset{NH_2}{\overset{|}{C}}H{-}COOH + H_2O \xrightarrow{\text{dithiomethylase}}$$

methionine

$$CH_3SH + NH_3 + CH_3{-}CH_2{-}\overset{O}{\overset{\|}{C}}{-}COOH$$

methyl-mercaptam *α-keto-butyrate*

Fig. 8.4. Oxidation of methionine.

1.2. Immobilization

Sulphur, often as sulphate, may be assimilated by microorganisms to produce sulphur amino acids. Sulphur accounts for some 0.1 to 1% of a microorganism's dry weight and deficiencies in plants, arising from this form of immobilization, are rare.

Reductive assimilation involves activation of sulphate using energy from ATP [I, II] and its subsequent reduction to sulphite [III] and then sulphide [IV] using NADPH.

$$SO_4^{2-} + ATP \xrightarrow{\text{ATP sulphurylase}} \text{adenosine phosphosulphate (APS)} + PPi \qquad [I]$$

$$APS + ATP \xrightarrow{\text{APS kinase}} \text{phosphoadenosine phosphosulphate(PAPS)} + ADP \qquad [II]$$

$$PAPS \xrightarrow[\text{NADP}]{\text{NADPH}} PAP + SO_3^{2-} \qquad \text{(PAPS reductase)} \qquad [III]$$

$$SO_3^{2-} \xrightarrow[\text{NADP}]{\text{NADPH}} S^{2-} + H_2O \qquad \text{(sulphite reductase)} \qquad [IV]$$

Hydrogen sulphide is converted to organic sulphur by condensation with serine (Fig. 8.5) and by other mechanisms.

$$\begin{matrix} H_2C{-}OH \\ | \\ HC{-}NH_2 \\ | \\ COOH \end{matrix} + H_2S \xrightarrow{\text{serine sulphydrase}} \begin{matrix} H_2C{-}SH \\ | \\ HC{-}NH_2 \\ | \\ COOH \end{matrix} + H_2O$$

serine *cysteine*

Fig. 8.5. Conversion of hydrogen sulphide to cysteine.

1.3. Oxidation

Oxidation of inorganic sulphur compounds may involve a shift of eight electrons (from −2 to +6). The variety of oxidation-reduction states (Fig. 8.6) means that a range of enzymic and chemical systems are involved in the transformation of reduced sulphur.

$$\underset{\textit{sulphide}}{H_2S} \longrightarrow \underset{\textit{sulphur}}{S} \longrightarrow \underset{\textit{thiosulphate}}{S_2O_3^{2-}} \longrightarrow \underset{\textit{tetrathionite}}{S_4O_6^{2-}} \longrightarrow \underset{\textit{trithionate}}{S_3O_6^{2-}} \longrightarrow \underset{\textit{sulphite}}{SO_3^{2-}} \longrightarrow \underset{\textit{sulphate}}{SO_4^{2-}}$$

Fig. 8.6. Oxidation-reduction states of sulphur.

The microorganisms catalyzing these changes fall into four broad categories.

1.3.1. Thiobacillus/Ferrobacillus Group. Members of these facultative or obligately autotrophic genera are typical pseudomonads

TABLE 8.6
Sulphur-oxidizing *Thiobacillus* species

Species	Substrates	Example of Oxidation
T. thioparus	$S_2O_3^{2-}$, $S_4O_6^{2-}$, $S_3O_6^{2-}$, H_2S	$5Na_2S_2O_3 + 4O_2 + H_2O \rightarrow 5Na_2SO_4 + H_2SO_4 + 4S$
T. thiooxidans	S^0, $S_2O_3^{2-}$	$2S + 3O_2 + 2H_2O \rightarrow 2H_2SO_4$
T. novellus	S, $S_2O_3^{2-}$	$Na_2S_2O_3 + 2O_2 + H_2O \rightarrow 2NaHSO_4$
T. ferrooxidans	S_2, S^0, $S_2O_3^{2-}$, Fe^{2+}	$2FeS_2 + 2H_2O + 7O_2 \rightarrow FeSO_4 + H_2SO_4$
T. denitrificans	S^0, H_2S, $S_2O_3^{2-}$	$5S + 6KNO_3 + 2H_2O \rightarrow K_2SO_4 + 4KHSO_4 + 3N_2$

(gram negative rods with a single polar flagellum). The dividing line between the two genera is indistinct and rests, to some extent, on the microbe's ability to oxidize sulphur and/or ferrous iron. Some of the characteristics and substrates of *Thiobacillus* species are summarized in Table 8.6.

The oxidation of sulphur or sulphide by *Thiobacillus* species involves their reaction with the sulphydryl group of glutathione to form a glutathione sulphide complex (glutathione polysulphide). Sulphide oxidase then converts sulphite to sulphate by a cytochrome-linked mechanism. Thiosulphate substrates are split into sulphite and sulphur. The sulphite is oxidized to sulphate and the other sulphur atom is converted into elemental sulphur and sulphides which may then proceed as before (Fig. 8.7).

$$\left.\begin{matrix} S^0 \\ S^{2-} \\ S_2O_3^{2-} \end{matrix}\right\} \longrightarrow R{-}S{-}S \longrightarrow SO_3^{2-} \xrightarrow[ATP]{ADP} SO_4^{2-}$$

Fig. 8.7. Oxidation of sulphur by *Thiobacillus* species.

1.3.2. Heterotrophs. A variety of heterotrophic bacteria, fungi and actinomycetes will oxidize elemental sulphur or thiosulphate in the presence of an organic substrate.

1.3.3. Trichome-Formers. A number of aquatic bacteria will oxidize hydrogen sulphide and deposit intracellular globules of elemental sulphur (Fig. 8.8). Examples of such bacteria are *Beggiatoa, Thiothrix, Thioplaca* and *Sphaerotilus.*

$$2H_2S + O_2 \longrightarrow 2S + 2H_2O$$

Fig. 8.8. Oxidation of hydrogen sulphide by filamentous bacteria.

1.3.4. Photosynthetic Sulphur Bacteria. Many photosynthetic bacteria perform the anaerobic oxidation of sulphur. Green sulphur bacteria, such as *Chlorobium*, may deposit sulphur intracellularly (Fig. 8.9). Purple sulphur bacteria (*Chromatium*) further oxidize the sulphur to sulphate (Fig. 8.10). This close coupling between photosynthesis and sulphur oxidation was first proposed in 1941.

$$2H_2S + CO_2 \longrightarrow (CH_2O) + H_2O + 2S$$

Fig. 8.9. Oxidation of sulphur by green sulphur bacteria.

$$H_2S + 2H_2O + 2CO_2 \longrightarrow 2(CH_2O) + H_2SO_4$$

Fig. 8.10. Oxidation of sulphur by purple sulphur bacteria.

1.4. Reduction

Soils deficient in oxygen have high numbers of bacteria which use oxidized forms of sulphur as electron acceptors. Wet paddy soils, for example, may contain greater than 1×10^6 sulphate reducing organisms per gram. The predominant bacteria involved in desulphuration are the vibriods belonging to the genus *Desulphovibrio* and, in thermophilic environments, *Clostridium* (*Desulphotomaculum*) *nigrificans*.

The dissimilatory reduction of sulphate to sulphide differs from the synthesis pathway described in 1.2. in that APS is reduced to sulphite and adenosine monophosphate (AMP) by APS reductase. The reductive process may be coupled to the heterotrophic oxidation of carbohydrates, organic acids (Fig. 8.11) or alcohols.

$$\underset{\textit{lactic acid}}{2CH_3\overset{\overset{\displaystyle OH}{|}}{C}H{-}COOH} + SO_4^{2-} \longrightarrow \underset{\textit{acetic acid}}{2CH_3{-}COOH} + 2CO_2 + H_2S + 2OH^-$$

Fig. 8.11. Bacterial reduction of sulphate.

2. Acid Drainage Water

The ferrous sulphide minerals, pyrite and marcasite, are frequently associated with coal seams and are left behind as waste products during mining operations. Oxidation of this reduced iron and sulphur occurs both chemically and microbiologically with the result that water draining from mine workings has a pH of between 1.5 and 4.0. Furthermore, soluble nutrients are frequently in high concentrations—especially sulphur, calcium, magnesium and iron. Microbial oxidation of ferrous sulphide is mainly performed by two acidophilic chemolithotrophs *Thiobacillus thiooxidans* and *Thiobacillus ferrooxidans*. A stalked bacterium, *Metallogenium*, is also implicated.

The chemical and biological transformations relevant to acid water production are summarized in Table 8.7. In the absence of moisture pyrittic ion is oxidized to ferrous sulphate and-sulphur dioxide (i). In the presence of water, ferrous sulphate and sulphuric acid are produced (ii). Ferrous sulphate may be further oxidized to ferric sulphate (iii) which is then hydrolysed to ferric hydroxide (iv). It is the mixture of precipitated ferric salts which gives the water draining from coal mines its characteristic yellow-gold colour.

TABLE 8.7

Transformations giving rise to mine water pollution

(i)	$FeS_2 + 3O_2 \rightarrow FeSO_4 + SO_2$
(ii)	$2FeS_2 + 7O_2 + 2H_2O \rightarrow 2FeSO_4 + 2H_2SO_4$
(iii)	$4FeSO_4 + 2H_2SO_4 + O_2 \rightarrow 2Fe_2(SO_4)_3 + 2H_2O$
(iv)	$2Fe_2(SO_4)_3 + 12H_2O \rightarrow 2Fe_2(OH)_6 + 6H_2SO_4$
(v)	$FeS_2 + Fe_2(SO_4)_3 \rightarrow 3FeSO_4 + 2S$
(vi)	$2S + 3O_2 + 2H_2O \rightarrow 2H_2SO_4$

Ferric sulphate is itself an oxidant and may contribute to pyrite transformations (v) and produce sulphuric acid as does the direct oxidation of sulphur (vi).

One of the ecological consequences of acidic effluents is to render the water courses into which they drain unsuitable for growth of aquatic plants, fish and many microorganisms. In addition, the water is unsatisfactory for domestic consumption or irrigation purposes. A variety of methods have been proposed and tested in an attempt to counteract acid production in coal-mines, for instance, the stimulation of sulphate reducers (e.g. *Desulphovibrio*) in an attempt to reprecipitate sulphur as sulphide. This has been achieved by promoting anaerobic (reducing) conditions following the addition of organic matter, either directly to abandoned mines or to lagoons into which the acid waste has been directed. Cellulytic microflora are very good at breaking down organic waste (and consuming oxygen) at acid pH's and thus cellulose amendments (sawdust, waste paper) are most useful. It may even be possible to couple domestic sewage disposal with acid mine treatment.

Other methods of attacking the problem have included the addition of specific antimicrobial agents, such as α-keto acids and carboxylic acids, to inhibit *Thiobacillus* activity (although acid production is not the total responsibility of bacteria), and the precipitation of sulphur by treating effluents with limestone to elevate the pH.

3. Microbial Corrosion

Sulphur bacteria are involved in the destruction of metal in three basic ways. Sulphur oxidizers produce sulphuric acid and sulphate reducers evolve hydrogen sulphide—both of which have a direct corrosive effect on metal surfaces. In addition, sulphate reducers, and other bacteria, may cause electrochemical decay.

To understand how electrochemical decay comes about it is

necessary to recall that metals are composed of lattices of positively charged metal ions balanced by negatively charged electrons. Under certain conditions both these particles are capable of dissociating and passing into solution.

When corrosion occurs the metal ions move into the water either to remain in solution or be precipitated as hydroxide, sulphate, carbonate or chloride salts. This then leaves an abundance of negative charges in the metal, halting further positive ion diffusion unless the electrons are removed. The electron acceptors are oxygen or hydrogen (Fig. 8.12).

$$2H_2O + O_2 + 4e^- \longrightarrow 4OH^-$$

$$2H^+ + 2e^- \longrightarrow H_2$$

Fig. 8.12. Electron acceptors in metal corrosion.

The passage of metal ions into solution is described as the anodic process and the sites from which it occurs, the anodes. By contrast, the acceptance of electrons by some reducible material (O_2, H^+) is the cathodic reaction and the site of acceptance the cathode. On the metal surface, anode and cathode sites may be only a few molecules apart. The probability of a particular metal undergoing corrosion in this manner is related to its electrochemical potential and its environment. Those metals and alloys with high electrochemical potentials (platinum, gold, silver) do not allow metal ions to pass freely through their lattices and into aqueous solution. Those with low electrochemical potentials (zinc, aluminium, steel, iron) are more reactive and consequently more subject to corrosion.

It is in anaerobic environments that sulphate reducers (and other anaerobes) contribute to metal decay by consuming hydrogen ions and electrons, generated at the cathode, in the reduction of sulphate to sulphide (Fig. 8.13).

$$SO_4^{2-} + 8H^+ + 8e^- \longrightarrow S^{2-} + 4H_2O$$

Fig. 8.13. Reduction of sulphate.

The sulphide precipitates the Fe^{2+} produced at the anode as iron sulphide. The net result of these two reactions is that the equilibrium of iron dissociation (Fig. 8.14) is to the right and the process of corrosion (movement of Fe^{2+} into solution) is accelerated.

$$Fe \longrightarrow Fe^{2+} + 2e^-$$

Fig. 8.14. Dissociation of iron.

$$4H_2 + NO_3 \longrightarrow NH_3 + 2H_2O + OH^-$$

Fig. 8.15. Nitrate reduction by *Micrococcus denitrificans.*

It is apparent that those microbes which catalyse the oxidation of hydrogen ions, that is those producing hydrogenase, are key anaerobes in metal corrosion. Thus, although sulphate reducers have been and are still traditionally linked with the process other microbes, such as *Micrococcus denitrificans* (Fig. 8.15) and *Methanobacterium omelianskii* (Fig. 8.16) are also involved.

$$4H_2 + CO_2 \longrightarrow CH_4 + 2H_2O$$

Fig. 8.16. Carbon dioxide reduction by *Methanobacterium omelianskii.*

Electro-chemical corrosion can be controlled by placing pipes in well-drained, alkaline soils (low sulphate levels), protecting the metal surface with a coating of bitumen or polyvinylchloride tape or by galvanizing whereby the zinc coat will act as a "sacrificial" anode and be corroded preferentially.

RADIOACTIVE POLLUTION

1. Introduction

Organisms have always been exposed to low levels of radiation. Present day background levels result in an average annual dose to each human being of about 0.1 rad (the rad is a unit of incident ionizing radiation equivalent to 100 erg gm^{-1}). Natural isotopes contribute 50% of this dose, cosmic rays 25% and irradiation from components of the cytoplasm, the remainder. In the last thirty years there has been not only an increase in the absolute amounts of radioisotopes in the environment, but also the introduction of new ones. This has reflected the development and testing of nuclear weapons and the use of isotopes in industry, research and medicine. In the U.S.A. alone more than 1000 megacuries of radioactive waste are produced per year (1 curie = 3.7×10^{10} disintegrations sec^{-1}). The increase in total radioactivity in the environment, as a result of these recent activities, is very small—about two per cent (Table 8.8).

Radioactive pollution may simply be described as that level of radiation that produces harmful effects, but it is extremely difficult to define that level quantitatively. It may be that any radiation, even the background level, is responsible for some disease amongst higher

TABLE 8.8

Present day exposure of Man to radioactivity

Radiation source	Annual dose (rads)
Natural background from:	
cosmic rays	0.025
rocks	0.050
within the body	0.025
Total Natural	0.100
Fall-out from weapons testing	0.0013
Waste disposal (nuclear power, industry, medicine, etc.)	0.0003–0.003
Total Nuclear Industry	0.0016–0.0043

organisms. This hypothesis is difficult to prove or disprove but in view of this uncertainty it is desirable to keep any further increases to a minimum. If Man is to use radioisotopes with complete safety, it is necessary to determine precisely the limits of exposure, the nature of interactions between radiation and cytoplasm and how much any biological damage can be prevented or repaired.

Accidents involving radioisotopes are rare, no doubt due to the elaborate precautions taken in handling large amounts of radioactive material. However, in a few cases accidents have caused serious illness or even death, even though the amounts of material released have not been significant on a world scale. Widespread exposure to high radiation levels is a rare phenomenon (Hiroshima and Nagasaki were

TABLE 8.9

Some hazardous radioisotopes

Isotope	Half-life	Principal radiation	Maximum permissible amounts (MPA) for continuous total body exposure (μC)
Plutonium-239	24 400 years	α	0.04
Radium-226	1 620 years	α	0.10
Polonium-210	138 years	α	0.04
Strontium-90	19.9 years	β	1.0
Calcium-45	152 days	β	13.0
Carbon-14	5 600 years	β	2.5
Phosphorus-32	14.3 days	β	10.0
Iron-59	45.1 days	β	11.0
Iodine-131	8.1 days	β	0.3

exceptions), but Man's nuclear activities have resulted in the world-wide distribution of several isotopes that are especially dangerous when absorbed into the body (Table 8.9). A number of these are characterized by both long radioactive and biological half-lives, and may accumulate in the liver, spleen, thyroid and other areas. Some, such as strontium-90, plutonium-239 and radium-226, have properties that are similar to calcium and, as a result, become concentrated in skeletal tissue.

In addition, some isotopes tend to accumulate in limited areas of tissues ("hot spots") which renders them even more dangerous. Attempts have been made to establish the maximum permissible amount (MPA) for each radioisotope in the body, the atmosphere and natural waters. The margin of safety in these figures is difficult to assess. For example the MPA for radium-226 in the skeleton has been estimated as 0.1 μC because there is no evidence that an individual with this amount of radium has suffered any harm. However, it is difficult to be certain about this because there are considerable variations in individual susceptibility.

2. Distribution of Radioactivity in the Environment

The distribution of natural radioisotopes and fission products in the environment is an important consideration, especially with respect to their accumulation in food chains. The major contributors to background radiation are uranium-238, thorium-232 and potassium-40. These are all easily oxidized metals, their oxides having low densities and so are found chiefly in the earth's crust. Most of the radioactivity that reaches living organisms originates from less than 20-cm below the ground. Radon (a radioactive gas) slowly diffuses from the soil into the air where its breakdown products are the main sources of radioactivity. As rocks weather, radioactive components become dissolved in the sea. Both thorium and uranium subsequently form insoluble salts that are precipitated onto the sea bed. Radioactive potassium is also precipitated, not because of salt formation but because it becomes adsorbed onto clay and organic particles. Uranium, thorium and most of their breakdown products do not enter food chains. Some, however, such as radium-226 and lead-210 (formed from uranium) and radium-228 (formed from thorium) are taken up by food plants, and may eventually become concentrated in mammalian bones.

Potassium-40 is the major source of radioactivity in protoplasm whilst carbon-14, another natural component of protoplasm, is derived from radioactive carbon dioxide originally formed by the influence of cosmic rays on nitrogen in the upper atmosphere. It has been calculated that the "average carbon dioxide molecule" spends five years in the atmosphere, five years in the surface layers of the oceans and 1200 years in the deep waters. Organisms consist almost entirely of carbon compounds and water and so the entry of carbon-14 into living tissues is a common event.

The distribution and ultimate fate of fall-out products from nuclear explosions varies with the magnitude of the detonation. Low to medium yield explosions (20-200 kilotonnes) force fission products up into the trophosphere (turbulent lower layers of the atmosphere, 0-10 km) from which fall-out is quite rapid (within a few months). Trophosphere winds do not generally cross the equator and so the products from such explosions are isolated to the hemisphere in which they occur. A proportion of the radioactive products of high yield nuclear explosions (greater than 200 kilotonnes) reach the stratosphere which transports them world-wide. The resulting fall-out is more widespread and occurs over many years.

Plants absorb more strontium-90 (and cesium-137) than do animals. As a result, strontium is diluted as its passes along a food chain (e.g. plants-cow-man). The strontium to calcium ratio in the plant depends on a variety of factors such as soil-type, nearby vegetation, method of cultivation and the species. The concentration of isotope in the milk of cows feeding on contaminated herbage decreases by about 90% and further dilution occurs after consumption by man. High strontium levels in humans are associated with vegetarian diets and there is a subsequent association between body strontium and the incidence of leukemia.

A large proportion of the highly radioactive waste resulting from the peaceful uses of atomic energy have been carefully contained. Nevertheless, a thorough understanding of the distribution and fate of these substances is required, firstly to establish the limits for the safe release of nuclear wastes and secondly to facilitate handling of any accidental spillage in excess of this level.

3. Management of Radioactive Wastes

The main sources of radioactive wastes are shown in Table 8.10.

There are three fates of radioactive wastes: dispersal in the environment, permanent confinement in a storage vessel and a combination of these two, preliminary storage to allow decay of unstable nuclides before release into the environment. Most waste is

TABLE 8.10

Major sources of radioactive wastes

Uranium mining and milling
Processing of uranium concentrate
Fuel fabrication
Reactor wastes
Fuel processing
Research and development laboratories
Hospital and biological laboratories
Isotope production plants
Nuclear powered transport

contained permanently in storage tanks (in the U.S.A., 99.9% of the total material).

Active material may be gaseous, liquid or solid. Gaseous effluents are seldom highly radioactive, but when they are they may be removed by filters, spraying with water, adsorption (onto activated charcoal or silver or copper mesh) or by using electrostatic collectors.

Large quantities of liquid wastes are generated and there are strict international regulations concerning disposal. Low level waste (less than 1 μCi ml^{-1}) can be released in a controlled fashion. Sewage treatment plants are quite effective in removing some nuclides (P-32, I-131, and Sr-90) from water. Medium level waste (1 μCi-10 mCi ml^{-1}) is usually concentrated by ion exchange, chemical precipitation or evaporation before treatment as high level waste (greater than 10 mCi ml^{-1}). Vermiculite, lignite, and sawdust are sometimes employed to absorb the liquid and give a manageable solid. There is already far too much high level liquid waste to disperse in the oceans and so most of this material is stored in carefully designed tanks and placed under constant vigilance. There is of course, a real danger of material being released during earthquakes, military conflict, etc. but a development which may overcome this problem involves fusion of the liquid into glasses (glassification), micas or ceramics before burial at sea or underground.

Some low level liquid wastes are discharged into the ground well above the water table. In this way the activity of those isotopes with short half-lives is reduced by decay before the entry of nuclides into ground water. Large amounts of material are also released into rivers and into the sea. Sea dispersal is rapid and effective for low activity waste, for example 1000 Ci released in one place on the surface would become spread over about 40 000 km^2 at an average concentration of about 1.5×10^{-10} μCi ml^{-1} after forty days. In setting maximum limits for the amount of radioactivity in water, its subsequent use is an important consideration and if it is to be human consumption it is necessary to set low limits. Cooling waters from nuclear power plants often flow into rivers but do not usually increase background levels of radiation.

Solid radioactive wastes are disposed of in several ways. Those which are slightly active (less than 0.1 $\mu Ci\ ml^{-1}$) are often incinerated whilst low level wastes are sometimes stored in concrete vessels above ground. High activity material may be sealed in containers and dropped into deep parts of the oceans. Large amounts are disposed of in the ground, low level waste being buried, but more active material is first enclosed in concrete containers. Deep mines and cavities in rock and salt have also been used.

4. Effect of Radiation on Microorganisms

Although, at the molecular level, the effect of radiation on microorganisms is similar to its effect on plants and animals, microbes are able to tolerate much higher levels. The reason for this is unclear but there is some correlation between sensitivity to ionizing radiation and degree of biological complexity. The Ld_{50} for man is 400-600 rads whilst that for most microbes is 1-50 krads although in the remarkably tolerant *Micrococcus radiodurans* the Ld_{50} is approximately 700 krads.

Radiation levels, even from freak accidents, are unlikely to be high enough to affect microbial populations in the biosphere. Nevertheless, a proportion of research has been concerned with the effects of the radiation on soil fungal and bacterial populations, enzyme activities, aggregate stability and nutrient availability.

Although soil microorganisms exhibit varying radio-sensitivities virtually all are killed by levels in excess of 2.5 Mrads. In contrast, extracellular enzyme activities (such as phosphatase and urease) appear somewhat more resistant. Microbial polysaccharides which cement sand, silt and clay components together may be depolymerized leading to disintegration of soil aggregates. In contrast the solubilization and release of nutrients (especially nitrogen and phosphorus) from soil humus and lysed microbial cells may even stimulate plant growth in radiation sterilized soils.

INDUSTRIAL POLLUTION

There are many forms of industrial pollution (Table 8.11). The offending effluents contain inorganic substances such as sulphides,

TABLE 8.11
Major industrial wastes

Industry	Origin of waste	Waste components and characteristics
FOOD		
Canning	fruit and vegetable preparation	colloidal, dissolved organic matter, suspended solids
Dairy	whole milk dilutions, buttermilk	dissolved organic matter (protein, fat, lactose)
Brewing, distilling	grain, distillation	dissolved organics, nitrogen fermented starches
Meat, poultry	slaughtering, rendering of bones and fats, plucking	dissolved organics, blood, proteins, fats, feathers
Sugar beet	handling juices, condensates	dissolved sugar and protein,
Yeast	yeast filtration	solid organics
Pickles	lime water, seeds, syrup	suspended solids, dissolved organics, variable pH
Coffee	pulping and fermenting beans	suspended solids
Fish	pressed fish, wash water	organic solids, odour
Rice	soaking, cooking, washing	suspended and dissolved carbohydrates
Soft drinks	cleaning, spillage, bottle washing	suspended solids, low pH
PHARMACEUTICAL		
Antibiotics	mycelium, filtrate, washing	suspended and dissolved organics
CLOTHING		
Textiles	desizing of fabric	suspended solids, dyes, alkaline, hot
Leather	cleaning, soaking, bating	solids, sulphite, chromium, lime, sodium chloride
Laundry	washing fabrics	turbid, alkaline, organic solids
CHEMICAL		
Acids	wash waters, spillage	low pH
Detergents	purifying surfactants	surfactants
Starch	evaporation, washing, etc.	starch
Explosives	purifying and washing TNT, cartridges	TNT, organic acids, alcohol, acid, oil soaps
Insecticides	washing, purification	organics, benzene, acid highly toxic
Phosphate	washing, condenser wastes	suspended solids, phosphorus, silica, fluoride, clays, oils, low pH

TABLE 8.11—*continued*

Industry	Origin of waste	Waste components and characteristics
Formaldehyde	residues from synthetic resin production and dyeing synthetic fibres	formaldehyde
MATERIALS		
Pulp and paper	refining, washing, screening of pulp	high solids, extremes of pH
Photographic products	spent developer and fixer	organic and inorganic reducing agents, alkaline
Steel	coking, washing blast furnace flue gases	acid, cyanogen, phenol, coke, oil
Metal plating	cleaning and plating	metals, acid
Iron foundry	various discharges	sand, clay, coal
Oil	drilling, refining	sodium chloride, sulphur, phenol, oil
Rubber	washing, extracting impurities	suspended solids, chloride, odour, variable pH
Glass	polishing, cleaning	suspended solids
ENERGY		
Steam power	cooling, coal drainage	dissolved solids, hot
Coal processing	cleaning, leaching of sulphur	high solids, sulphuric acid ferrous sulphate
Nuclear power	ore, fuel processing, cooling, water, laboratory wastes	acid, radioactive

cyanides, inorganic acids (steel industry), industrial organic compounds (such as carbon tetrachloride, phenols, pesticides and phenylalkanes), simple biochemicals (organic chemical, natural and synthetic fibre, and clothing industries) and biological material such as skin, fur, blood from the food industry and tanneries.

There are increasing moral and legal obligations to treat industrial pollution at source rather than contaminating the local environment or even relying upon municipal sewage treatment plants. Many industrial organics and most biological wastes can be treated microbiologically. In addition, there are a variety of procedures for dealing with the more toxic wastes. For example, there are four main ways of treating phenolic contaminants; (i) chemical oxidation with acidic potassium permanganate, (ii) adsorption into activated charcoal, (iii) extraction with isopropyl acetate or acetophenone followed by (i) or (ii), and (iv) biological treatment. The first two methods are adequate on most occasions whilst the third is used for high levels of pollutants and the last for concentrations of less than 300 ppm.

Inorganic pollutants pose a greater problem in that except at very low concentrations, biological treatment is generally excluded. In some cases the concentrations of salts in effluents are extremely high, as occurs in spent acid wastes from titanium oxide production which contain as much as 9% iron sulphate. Waste sulphuric acid from this and other sources is usually converted to sulphates or oxides which are then released into the sea.

THERMAL POLLUTION

1. Introduction

Thermal pollution may be defined as the output of heat by population and industry, such that the natural temperature of waters is increased; the natural temperature being loosely defined as that which would exist without an industrial society. The major contributor to thermal pollution is the power industry. Both steam- and nuclear-powered generating stations use large amounts of water for cooling. Of the two, nuclear installations produce more thermal pollution as they are less thermally efficient, and cooling water is usually returned to its source after a temperature increase of 6-9°C. There have been some extreme examples of thermal pollution in North America where waters have been heated from 15°C to 50°C or even, in a few cases to boiling point. Such increases in water temperature may have deleterious effects on the local fauna and flora by decreasing the concentration of dissolved oxygen which in turn increases their susceptibility to other pollutants. The deoxygenation effect only applies to pure waters although often water is extracted from polluted rivers in the first place. This means that the use of water for cooling purposes may actually have a beneficial effect on aquatic life because oxygen is introduced into the deoxygenated water as it passes down the cooling towers.

2. Ecological Considerations

Warm effluents are not automatically detrimental for, on occasion a change in the ecological balance of an aquatic ecosystem is desirable. However, in extreme conditions (i.e. greater than 60°C) only a few thermophilic bacteria may survive whilst above 40°C there may be a reduction in the diversity of plant, animal and microbial species or changes in the components of the population. For example, game fish (salmon, trout) tend to be replaced by coarse fish in high temperature waters. Where the cooling water is also chemically polluted, tubificid worms, able to live in polluted water, often reproduce more successfully at higher temperatures and the breeding season is extended. There have even been proposals to use local thermal pollution emitted from coastal power stations for sea fish farming.

It has been suggested that due to the increase in heat production, permanent climatic modifications may result. Although, on a local basis, changes have been recorded (the temperature in large cities may be raised above that of surrounding areas by as much as 3°C), the total heat emission from industrial processes is minute compared to the heat from the sun. Thermal pollution, at present levels, is unlikely to cause extensive climatic changes although atmospheric pollution may well do so (p. 217).

BIODETERIORATION

1. Introduction

A comprehensive treatment of biodeterioration is outside the scope of this book but several problems related to subjects discussed in earlier chapters, are described here. In addition, some important aspects of microbial biodeterioration are summarized in Table 8.12.

2. Deterioration of Cutting Fluids

Microbial degradation of cutting fluids was mentioned briefly in Chapter 4. There are three basic types of cutting fluid: straight oils, oil-water emulsions and soluble oils. Problems arising from microbial contamination are most serious with oil-water emulsions, although microbial spoilage does occur with other types as they do not remain water-free for long when in use. The main factors responsible for any deterioration of these oils are the poor quality of water used in preparation, poor hygeine around machinery, inadequate design of the feed systems and failure to use a suitable bacteriocide. Of these factors the quality of the water used is probably the most important. Water from a well or polluted river may introduce nutrients, that will promote microbial growth, or indeed microbial contaminants. Sometimes even tap water has definite disadvantages in that it usually contains sulphite which is reduced to sulphide in systems that tend towards anaerobiosis. Sulphide has a sickening odour and is corrosive, a characteristic enhanced by organic acids produced by fermentation processes. Also under anaerobic conditions, sulphide is involved in an electrochemical process in which it reacts with iron to

TABLE 8.12

Microbial processes involved in biodeterioration

Process	Examples
A. Material attacked chemically	
1. by microbes using it for food	a. Paper products attacked by cellulytic bacteria (*Cytophaga, Cellulomonas,* etc.)
	b. Food spoilage (*Penicillium, Mucor, Salmonella, Staphylococcus, Clostridium,* etc.)
	c. Milk souring (*Streptococcus, Lactobacillus*).
	d. Cutting-oil deterioration.
	e. Destruction of clothing and dwellings.
2. by the metabolic products of microorganisms	a. Inorganic acid production by *Thiobacillus* species (p. 220).
	b. Ammonia and hydrogen sulphide evolution.
	c. Organic acid production.
3. by electrochemical decay	Metal destruction by sulphate-reducers and others (p. 223).
B. Material or process is functionally impaired by the presence of microorganisms	a. Fuel lines and sewage filters (Chapter 3, p. 69) clogged by microbial biomass.
	b. Water pipes blocked by bacterial ferric and manganic oxide deposits.
C. Associated material attacked causing subsequent bio-deterioration.	a. Disruption of protective films.
	b. Degradation of synthetic polymer additives (Chapter 6, p. 181).
	c. Breakdown of corrosion inhibitors.

form ferrous sulphide. Hard tap water contains high calcium and magnesium levels which have a destabilizing effect on the emulsion and may also give rise to scale formation.

Biodeterioration is usually prevented or restricted by adding biocides or by sterilizing with irradiation (usually X-rays).

3. Deterioration of Plastics

The effects of microbes on plastics are discussed in detail in Chapter 6. All plasticized vinyl plastics are especially susceptible to microbial attack whilst fungi cause discolouration as well as destroying any cellulosic components that may be adjacent to the plastic. Bacterial contamination results in destruction of the plasticizer, causing the material to become stiff and brittle. It is usual to add long-lasting anti-microbial agents to such susceptible plastics.

4. Miscellaneous

Biodeterioration of foodstuffs has always presented economic and health problems, but the introduction of new preservatives and refrigeration have dramatically reduced the amount of spoilage.

The gas industry has had a rather special problem with the sulphate reducing bacteria, *Desulphovibrio desulphuricans* and *Clostridium* species which tend to accumulate at the bottom of water seals in gas holders, producing hydrogen sulphide which "spoils" the gas. *Thiobacillus thiooxidans* and *Thiobacillus concretivorus* generate sulphuric acid that can break down concrete and indeed on several occasions power station cooling towers have been rendered unsafe due to the activities of these bacteria.

Cosmetic and pharmaceutical products are also liable to spoilage, for example, anionic detergents in shampoos are degraded by various microorganisms, especially *Citrobacter* and *Aerobacter* species. Cosmetic creams are easily contaminated by pseudomonads which grow on the oil components.

A wide variety of microbes are responsible for problems in the paper industry and, in tanneries, microbial degradation of hides and skins occurs during the preliminary soaking process. Fungal growth may also cause spoilage during drying. Both bacteria and fungi are responsible for the biodeterioration of paint.

Recommended Reading

Air Pollution

Bach, W. (1972). "Atmospheric Pollution." McGraw-Hill.

McCormac, B. M. (Ed.) (1971). "Introduction to the Scientific Study of Atmospheric Pollution." D. Reidel Publishing Co., Dordrecht, Holland.

Singer, F. S. (1970). "Global Effects of Environmental Pollution", Parts I and II. D. Reidel Publishing Co., Dordrecht, Holland.

Williamson, S. J. (1973). "Fundamentals of Air Pollution." Addison-Wesley Publishing Co. Inc., Reading, Massachusetts, U.S.A.

Microbial Corrosion

Duggan, P. R. (1972). "Biochemical Ecology of Water Pollution", Chapter 11. Plenum Press.

Gall, J. Le. and Postgate, J. R. (1973). Physiology of sulphate reducing bacteria. *Advances in Microbial Physiology* 10, 81.

Holden, W. S. (Ed.) (1970). Corrosion and dissolution of metals by water, *In* "Water Treatment and Examination", Chap. 38. Churchill.

Miller, J. D. A. and Tiller, A. K. (1971). Microbial corrosion of buried and immersed metal. *In* "Microbial aspects of Metallurgy" (J. D. A. Miller Ed.), Chapter 3. Medical and Technical Publishing Co. Ltd, Aylesbury.

Trudinger, P. A. (1969). Assimilatory and dissimilatory metabolism of inorganic sulphur compounds by microorganisms. *Advances in Microbial Physiology*, 3, 111.

Radioactive Pollution

Arnold, J. R. and Martell, E. A. (1973). The circulation of Radioactive isotopes. *In* "Chemistry in the Environment". Readings from *Scientific American* 29, p. 272.
Mawson, C. A. (1965). "Management of Radioactive Wastes." Van Nostrand, Princeton.
Schubert, J. (1973). Radioactive poisons. *In* "Chemistry in the Environment". Readings from *Scientific American* 28, p. 267.

Industrial Pollution

Goronsson, B. (Ed.) (1970). "International Congress on Industrial Waste Water, Stockholm." Butterworths, London.
Gurnham, C. F. (Ed.) (1965). "Industrial Wastewater Control." Academic Press, New York.
Mulder, E. G., Antheunisse, J. and Crombach, W. H. J. (1971). Microbial aspects of pollution in the food and dairy industries. *In* "Microbial Aspects of Pollution" (G. Sykes and F. A. Skinner, Eds.). Academic Press, London.
Nemerow, N. L. (1963). "Theories and Practices of Industrial Waste Treatment." Addison-Wesley Publishing Co., Reading, Mass.

Biodeterioration

Erskin, N. A. M., Henderson, H. M. and Townsend, R. J. (1971). "Biochemistry of Foods." Academic Press, New York.
Walters, A. H. and Elphick, J. (Eds.) (1968). "Biodeterioration of Materials", Vol. 1. Elsevier, Amsterdam.
Walters, A. H. and Hueck-Van der Plas, E. H. (Eds.) (1972). "Biodeterioration of Materials", Vol. 2. Applied Science Publishers, Ltd., London.

Subject Index

Q

R

S